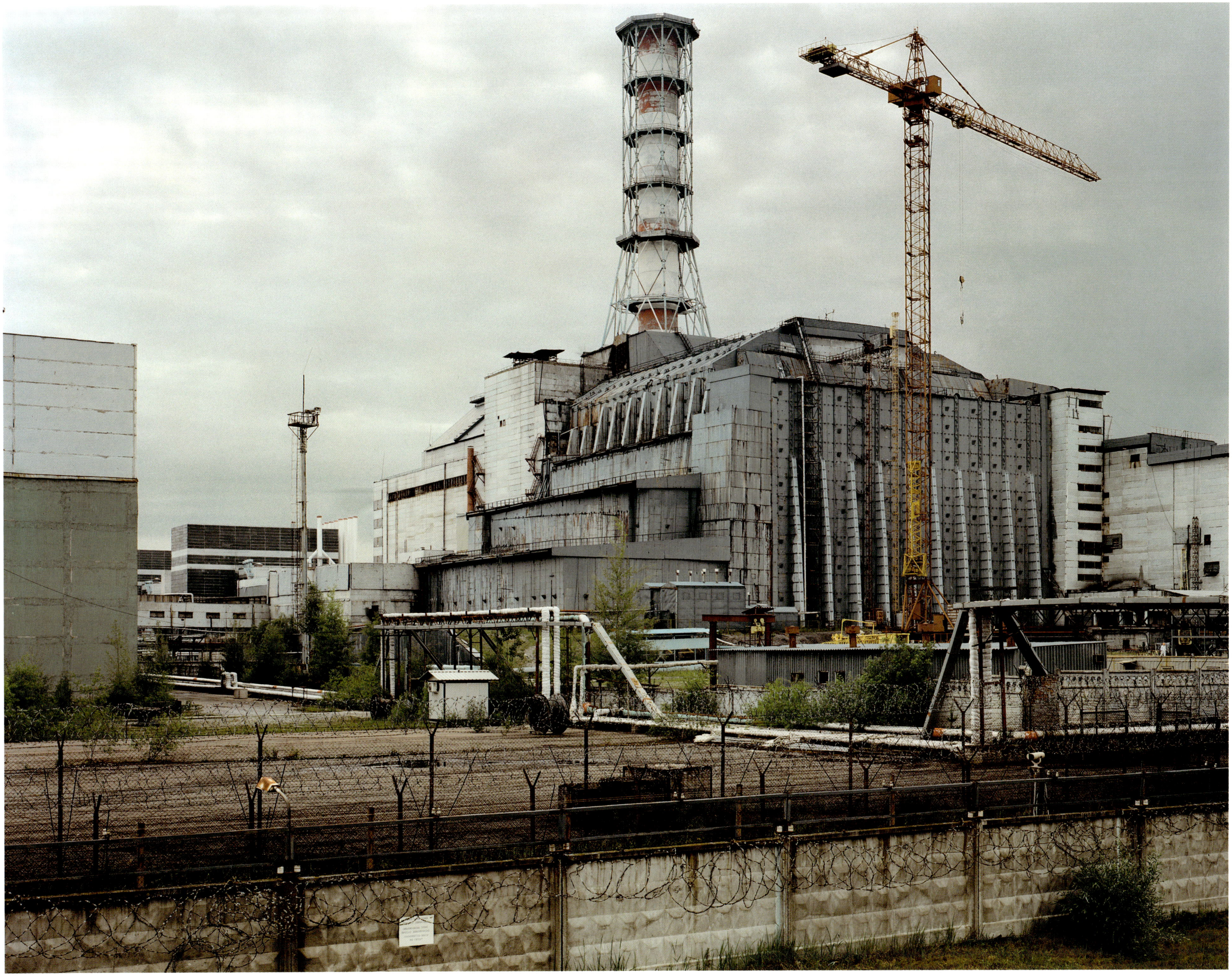

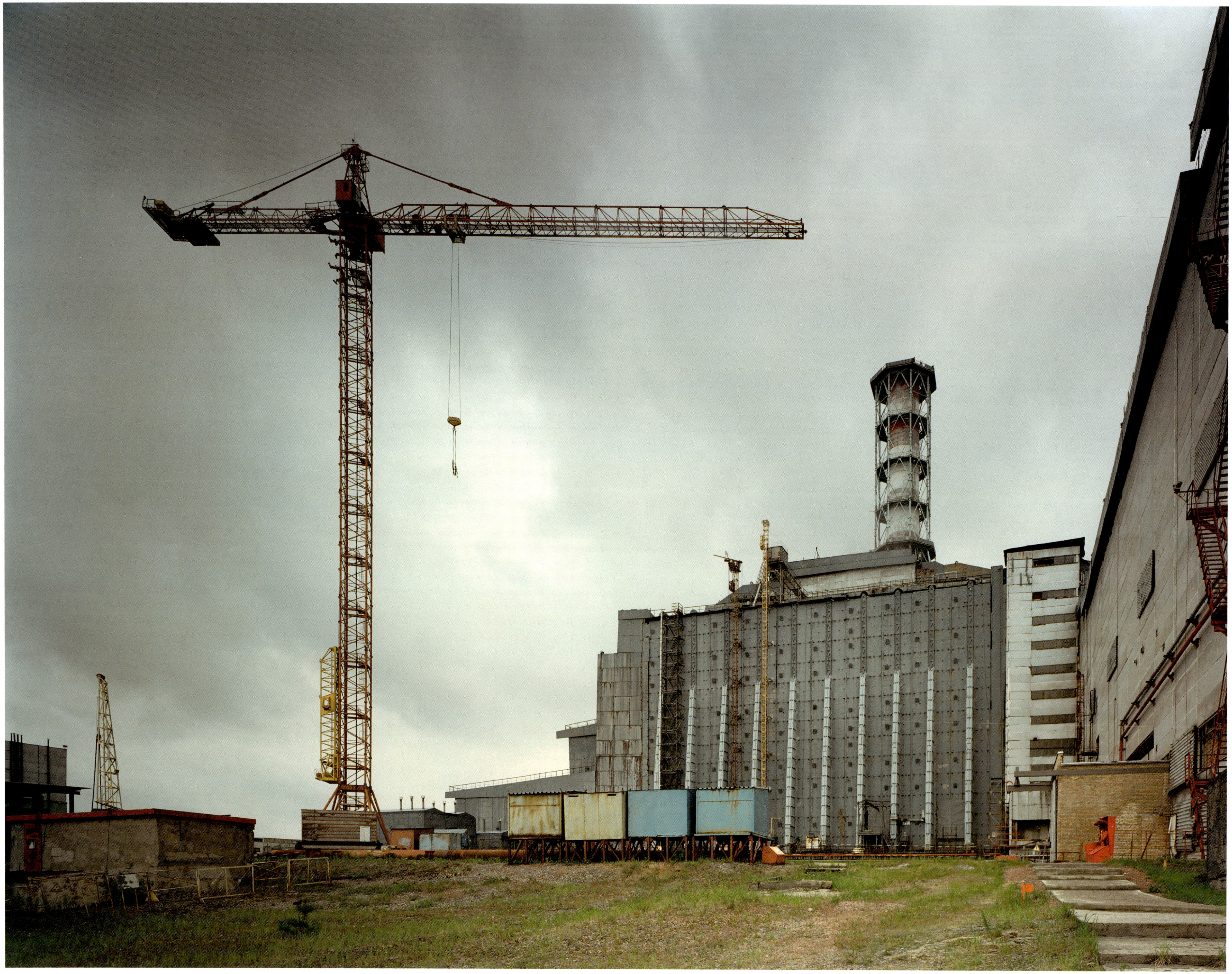

Robert Polidori

Zones of Exclusion

PRIPYAT AND CHERNOBYL

Text by Elizabeth Culbert

Steidl

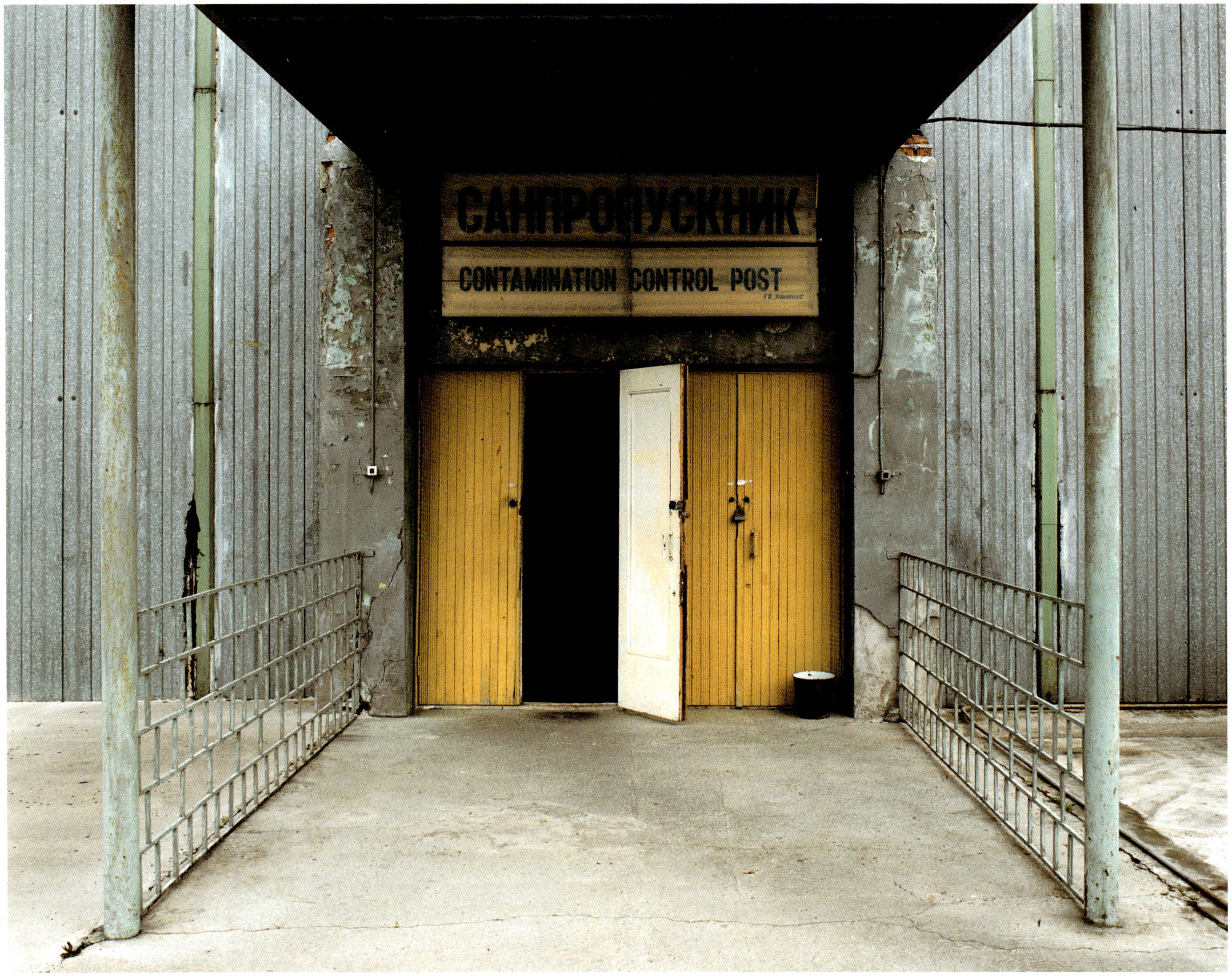
САНПРОПУСКНИК
CONTAMINATION CONTROL POST

"There's no return. Farewell. Pripyat, 28 April 1986."

Two days after explosions at the Chernobyl nuclear power plant released tons of radioactive material into the air, this message was written on the chalkboard of a nearby kindergarten classroom. It remains there today.

The residents of Pripyat, a community of plant workers, had been awakened in the early hours of April 26 to witness flames rising from Chernobyl into the night sky. A thousand kilometers away, the Kremlin received information before dawn about the catastrophe, but the people of Pripyat were given no indication of the dangers that faced them. When morning came, they carried on with the day's activities. A full 36 hours passed before an evacuation broadcast ordered them to leave their homes and their belongings behind. On that warm Saturday afternoon, families strolling through the streets were confronted by militiamen fully covered in protective clothing. Within minutes, they were part of an exodus of 50,000 people, rounded up by emergency workers and carried away by a fleet of buses to towns thought to be out of harm's reach.

It was a safety test that led to the destruction of Chernobyl's Unit 4 reactor and the worst nuclear accident in history. Engineers attempted to test the capability of the reactor's cooling system in the event of a power failure at the plant. Preparations for the test destabilized the reactor, but the engineer on duty, the shift foreman and the chief engineer chose to carry on. The reactor's emergency safety system was manually switched off.

Basic engineering deficiencies and the actions of the operators led to a massive steam explosion, followed by a rapid rise in reactivity. A second explosion blew the core apart and tore the roof off the building. Fuel, core parts and pieces of the structure landed on the ground outside and ignited multiple fires on the roofs of adjacent plant buildings. Large particles fell in the vicinity, but the bulk of the radioactive material rose into the atmosphere and was carried by the wind. It came down in rain over Ukraine, Belarus, Russia and much of Europe.

Mandatory evacuation continued over the next 10 days, forcing 116,000 people to depart from towns and villages within 30 kilometers of the plant—this area would be named the Exclusion Zone. Efforts to resettle the area's entire population were unsuccessful. Liquidators, as emergency workers came to be known, allowed Pripyat's residents to return in later weeks to gather their possessions. Many found their homes plundered; most of the objects they carried away were radioactively contaminated. Some people refused to leave, or returned after being forced out. Today the elderly are allowed to stay; children are not. No one is permitted to live within a 10-kilometer radius of Chernobyl. Pripyat remains abandoned.

Numerous waves of "absolute resettlement" continued through 1995. Nearly 350,000 people were moved from contaminated regions of Ukraine, Belarus and Russia, though hundreds of thousands still live in areas designated unsuitable for habitation—unsuitable not only because of radioactivity, but also because of severe socio-economic and psychological pressures. Poor and rural communities have no choice but to ignore the ban on farming and forestry; residents consume contaminated fish, milk and berries and burn radioactive wood in their stoves. The young and educated migrate at disproportionate rates. With limited local capacity to deal with health, economic and environmental challenges, living conditions spiral downward.

Late last year, thousands of Ukrainians—including pensioners and affected children—demonstrated in favor of reopening the Chernobyl plant (the last reactor was shut down in 2000). A partially functioning plant would provide jobs and power to the region. It would bring residents closer to having necessary food, medicine and rehabilitation.

Engineers, however, are concerned about the condition of the sarcophagus, the improvised concrete structure that liquidators hastily built around the damaged reactor building in the months following the catastrophe. Viewed as a temporary solution, the sarcophagus poses an increasing threat to safety; there are leaks and the ever-present threat of collapse. Structural reinforcements have been performed, but holes in the walls and ceiling continue to allow rain to enter and dust to escape. Birds and small animals pass in and out of the facility.

There are new plans to encase the sarcophagus in a massive steel shell that would buy scientists and engineers more time to find a way to handle its contents. Deep inside, core parts are buried randomly among tons of dust, fuel, concrete and combustibles. More than 200 tons of uranium and a ton of radioactive plutonium remain within.

Elizabeth Culbert
New York City, April 2003

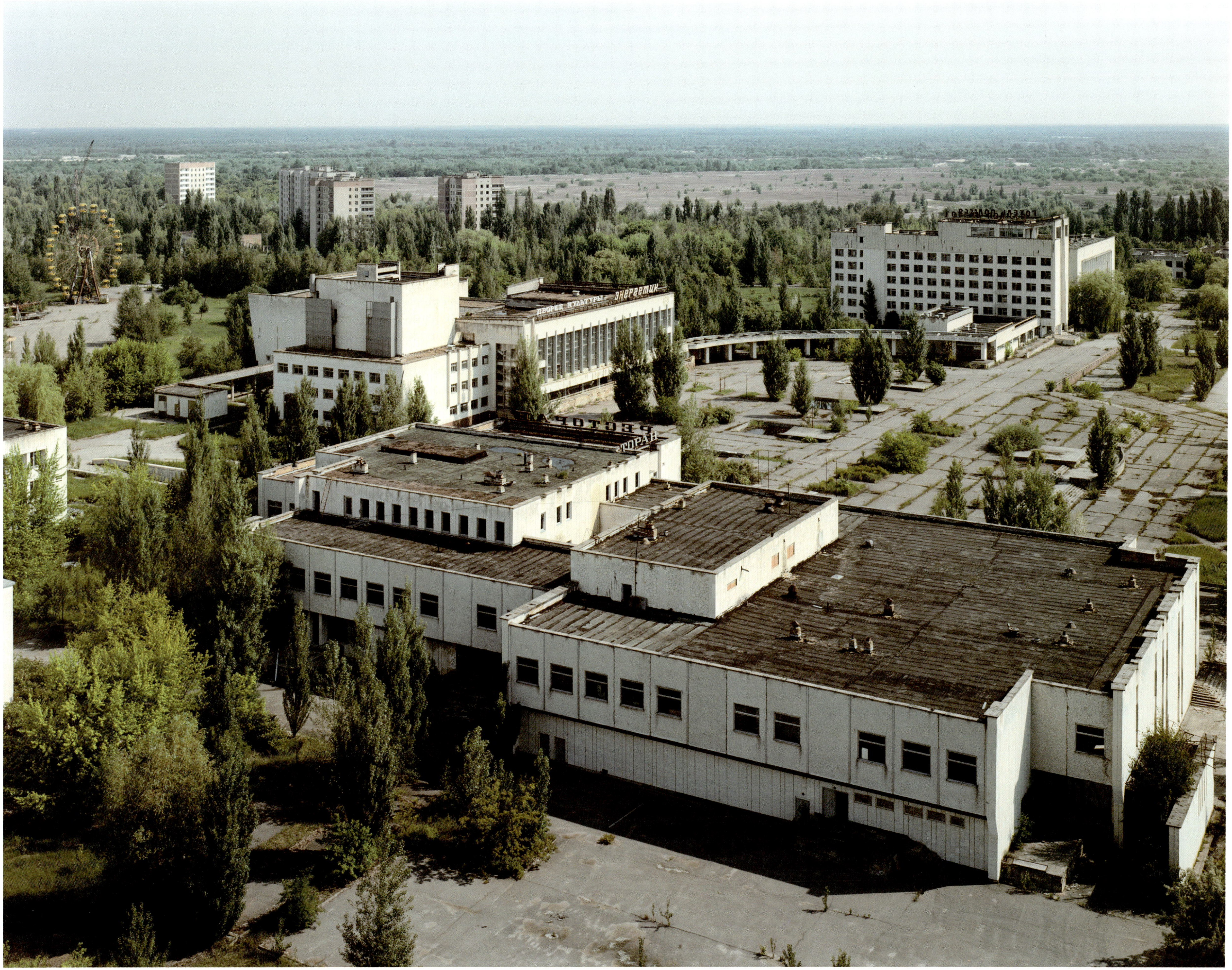
Дворец культуры Энергетик

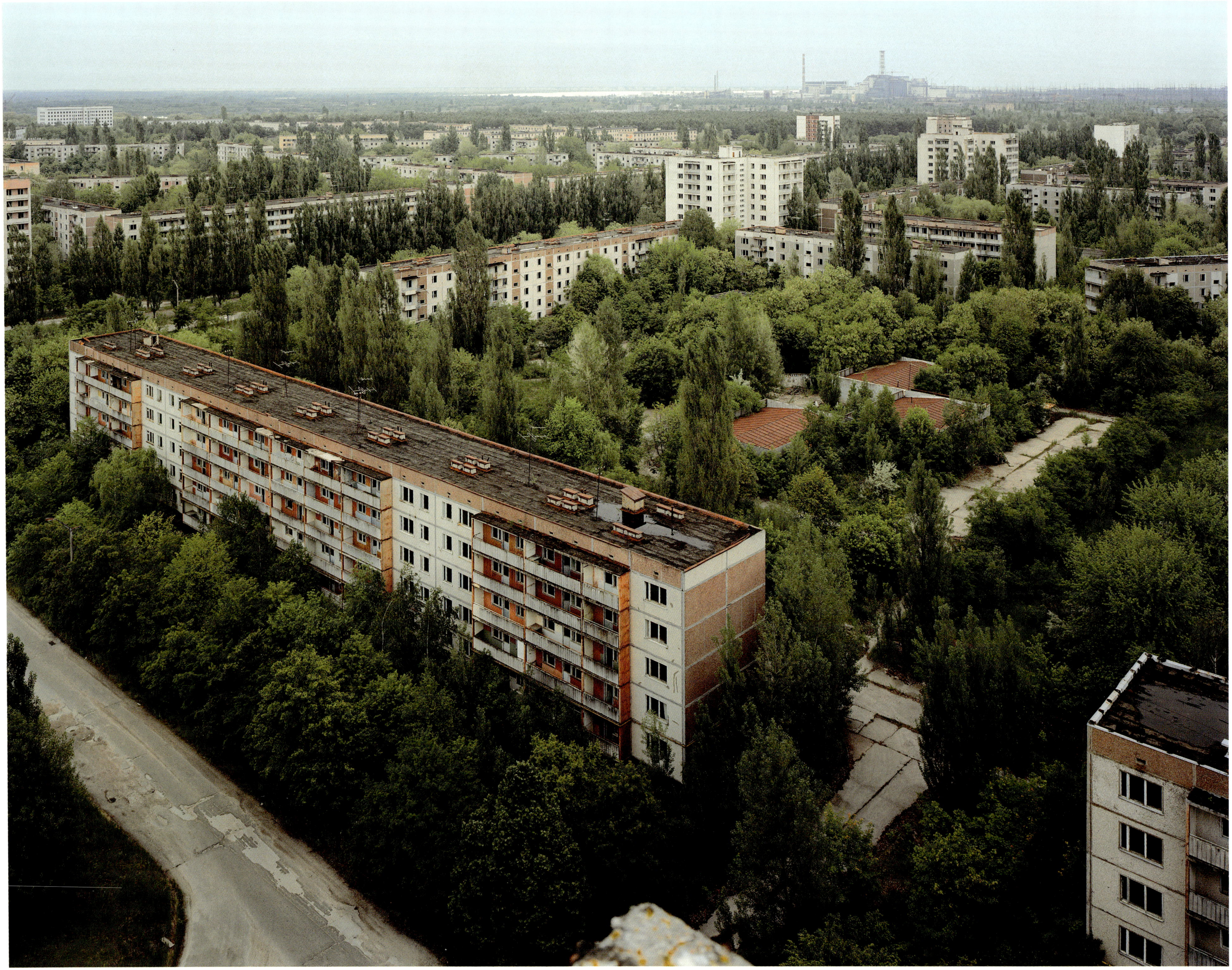

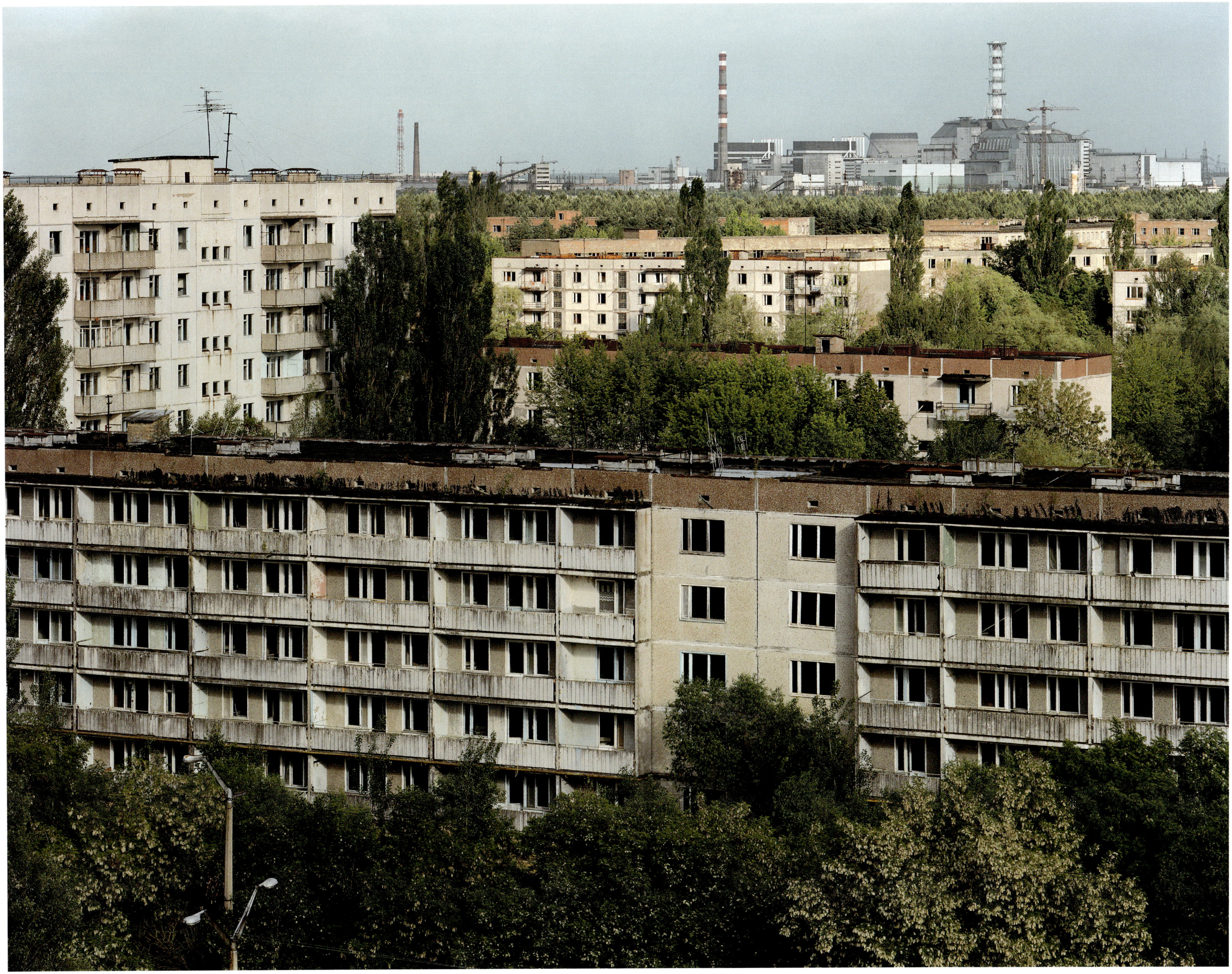

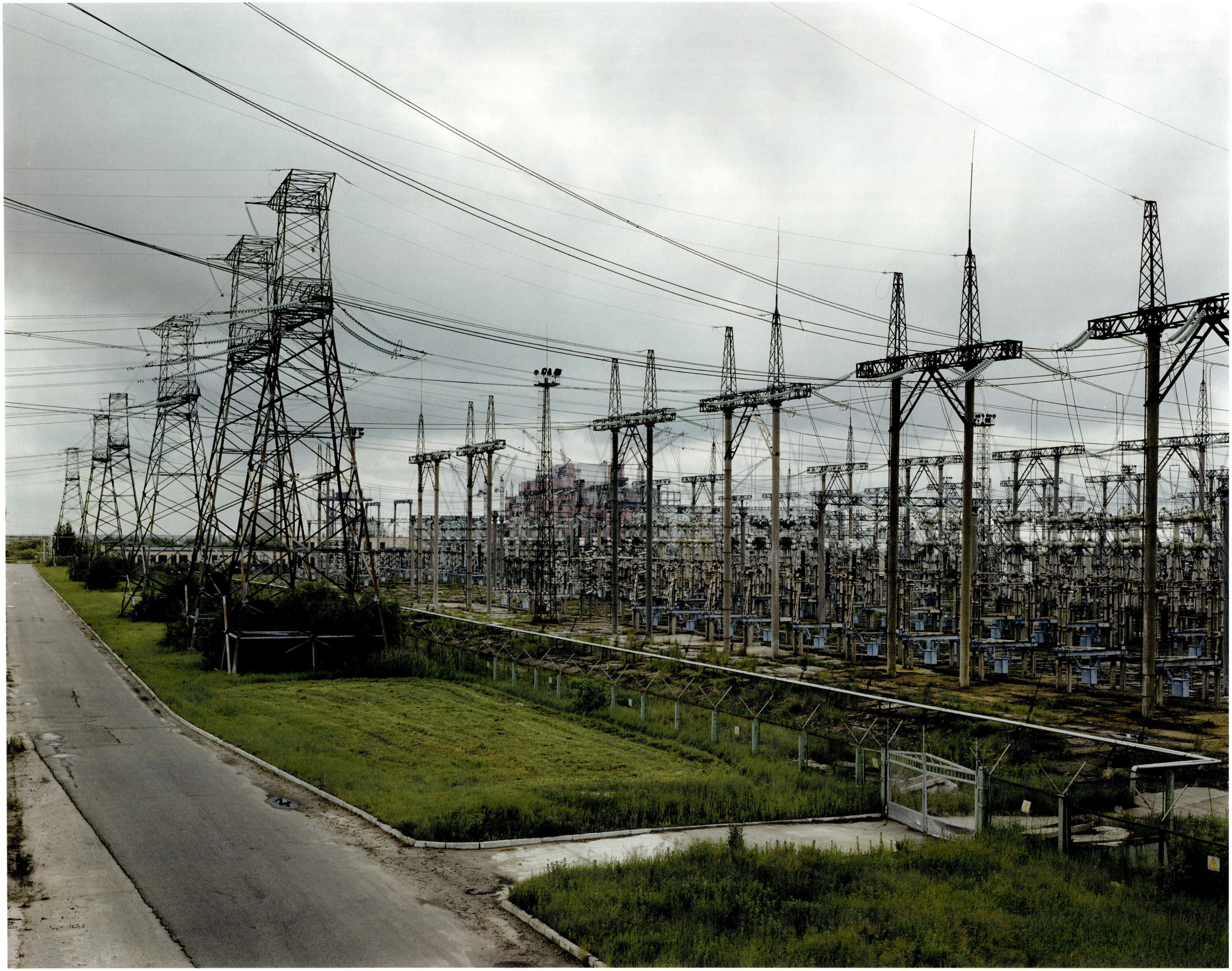

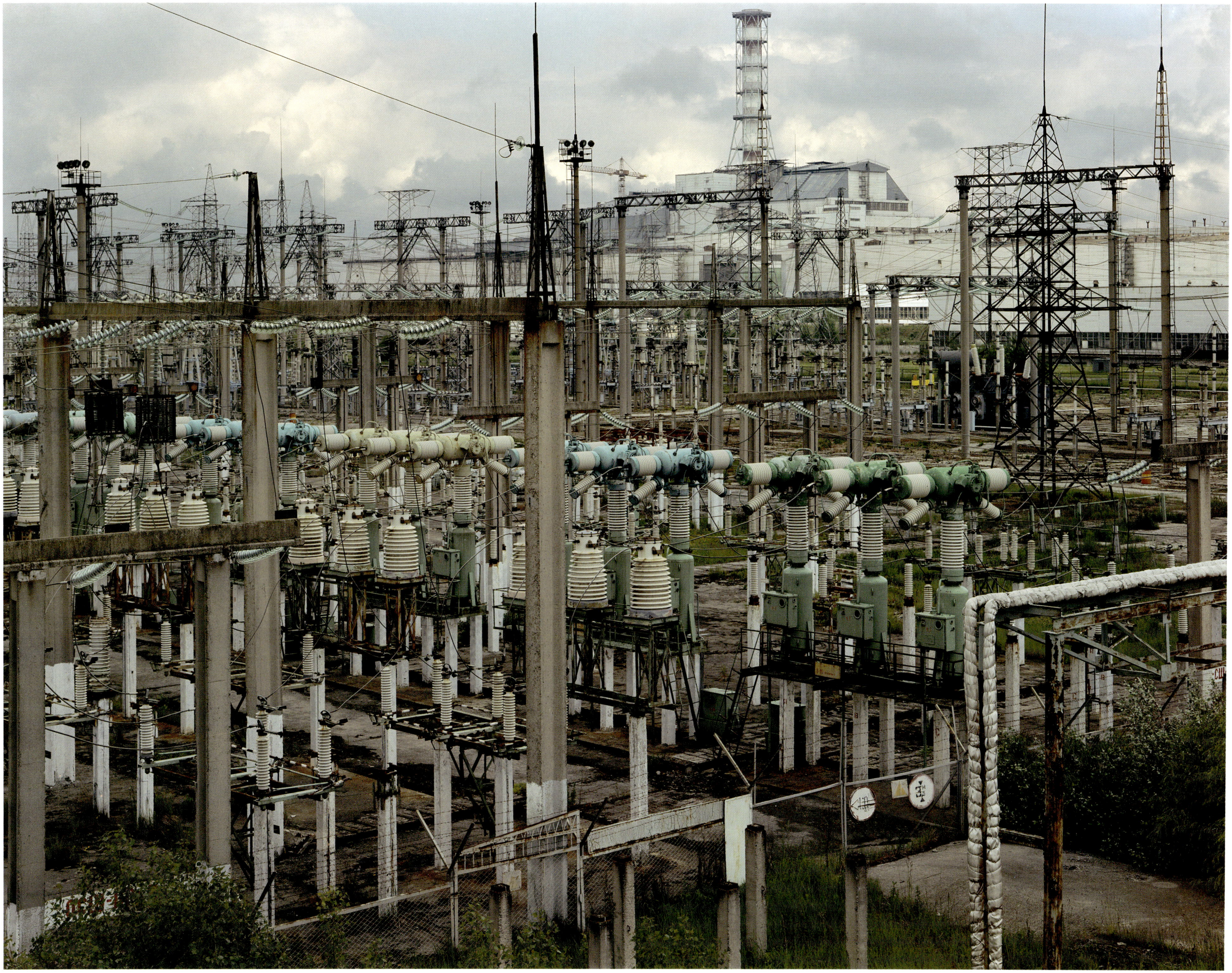

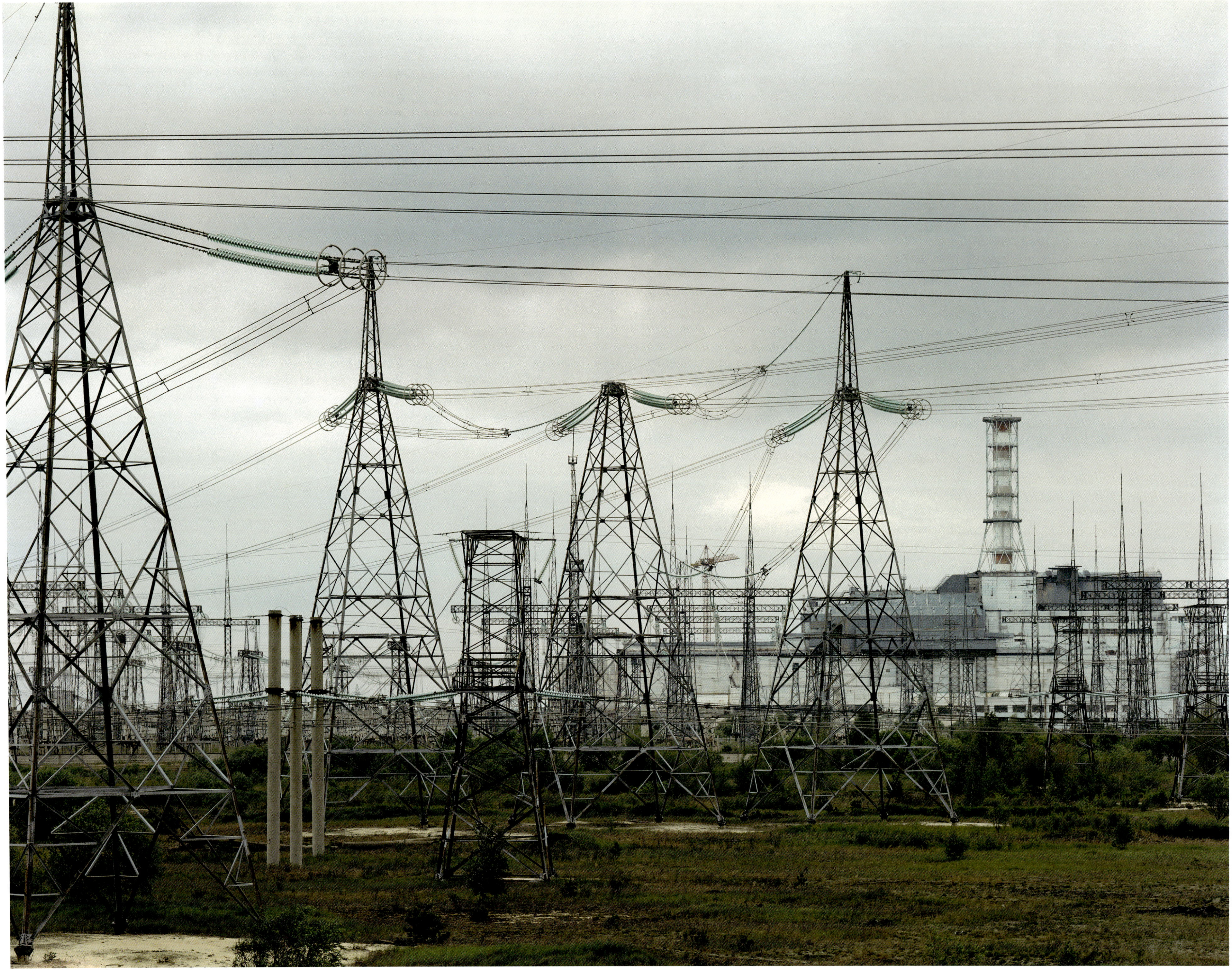

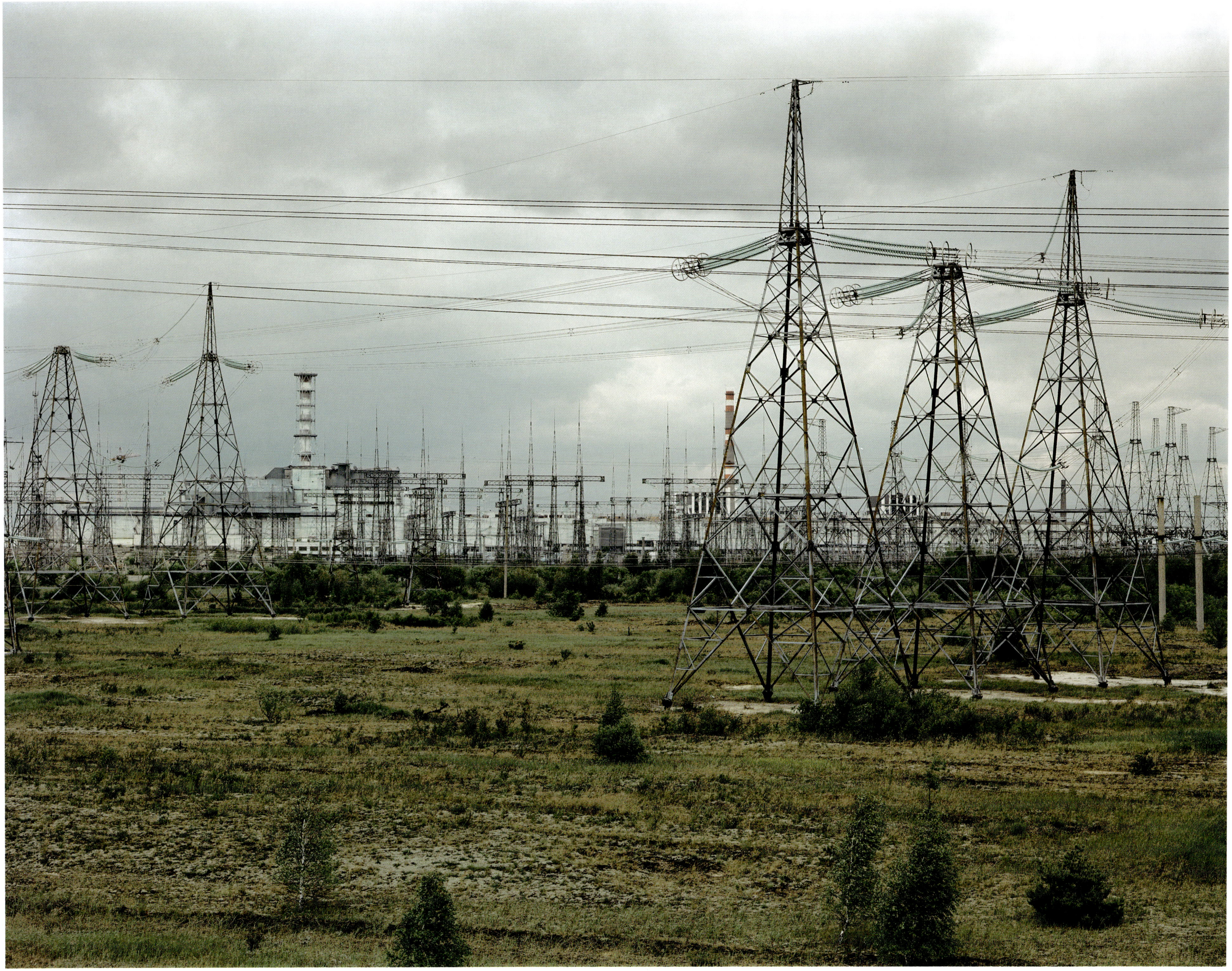

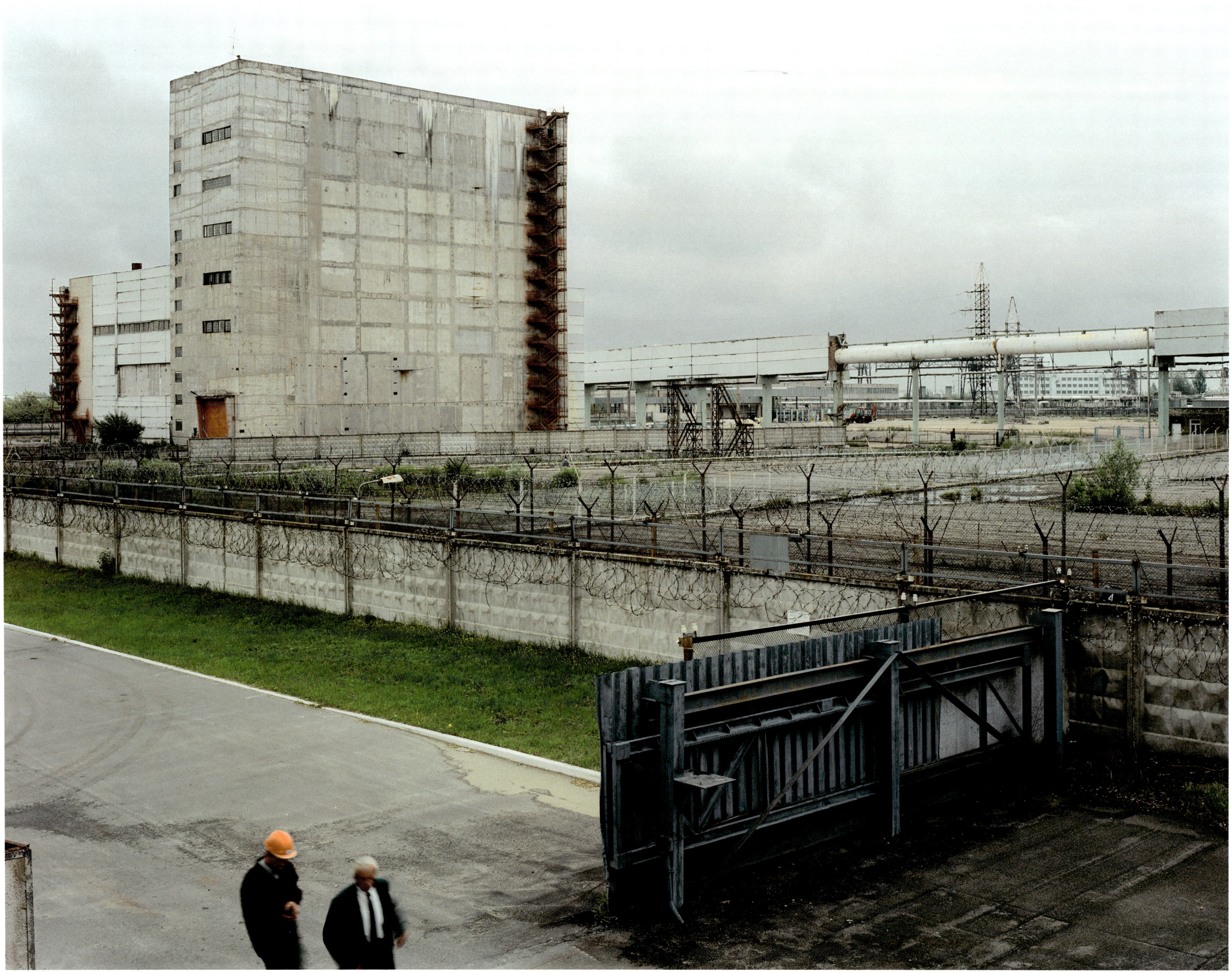

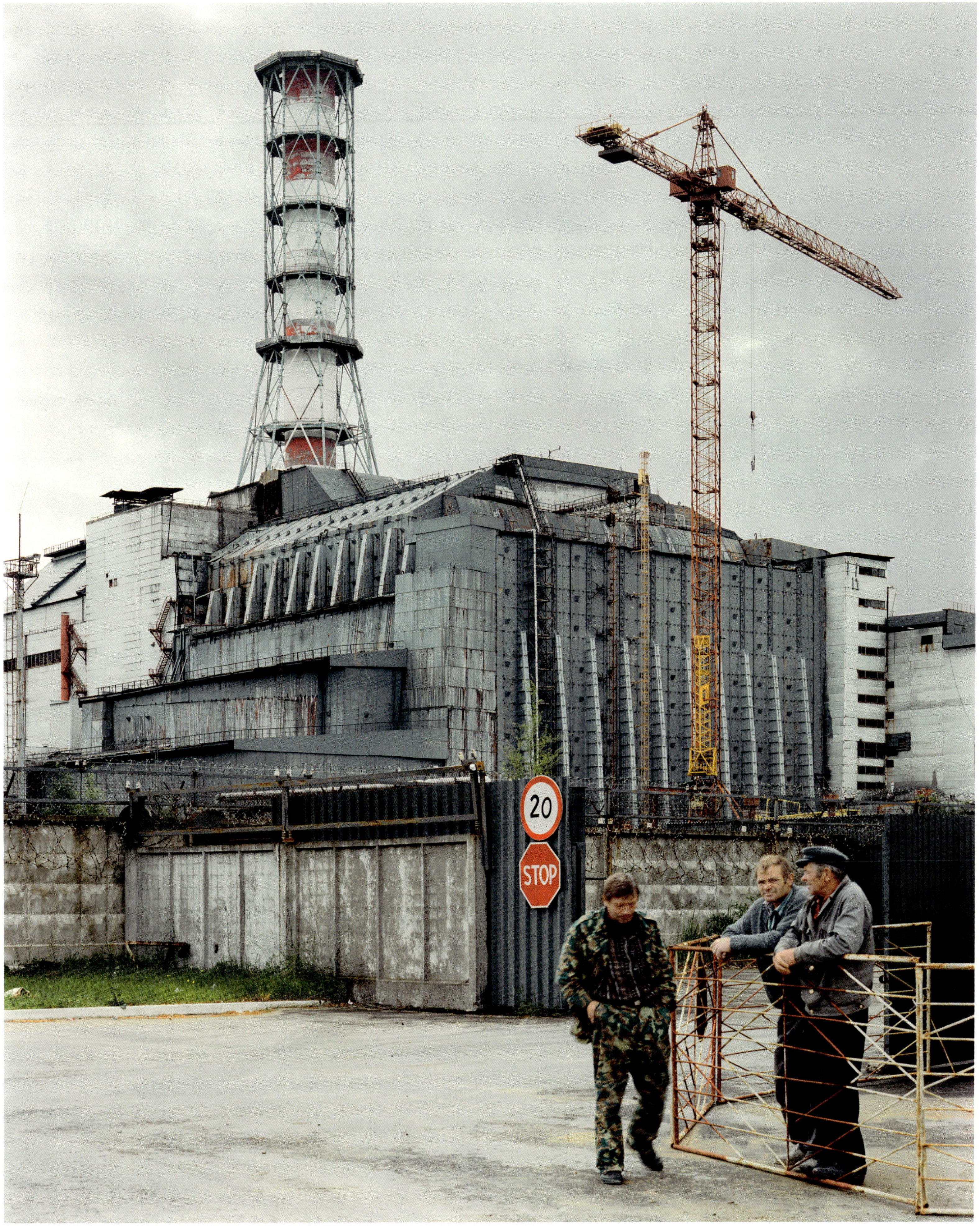
20
STOP

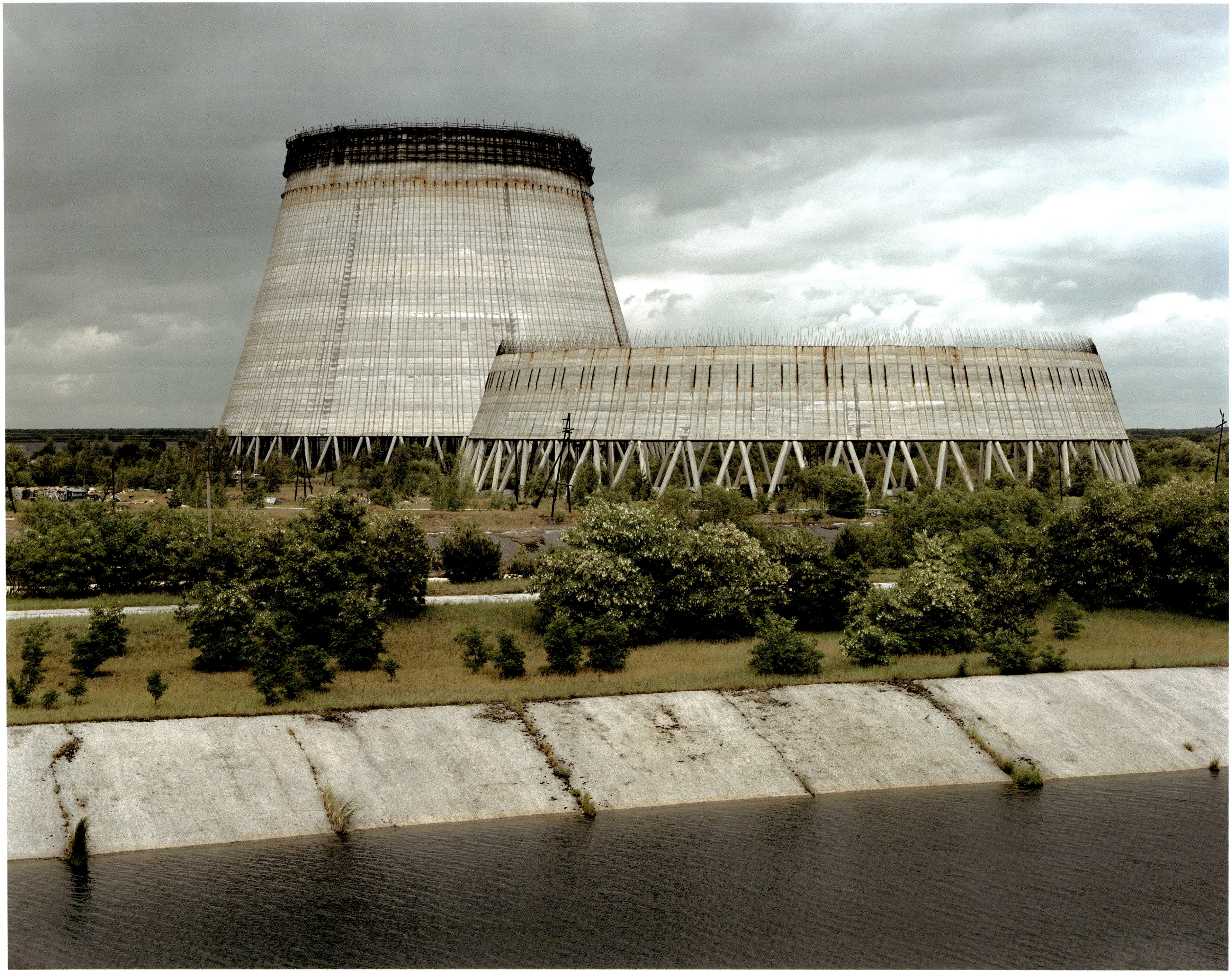

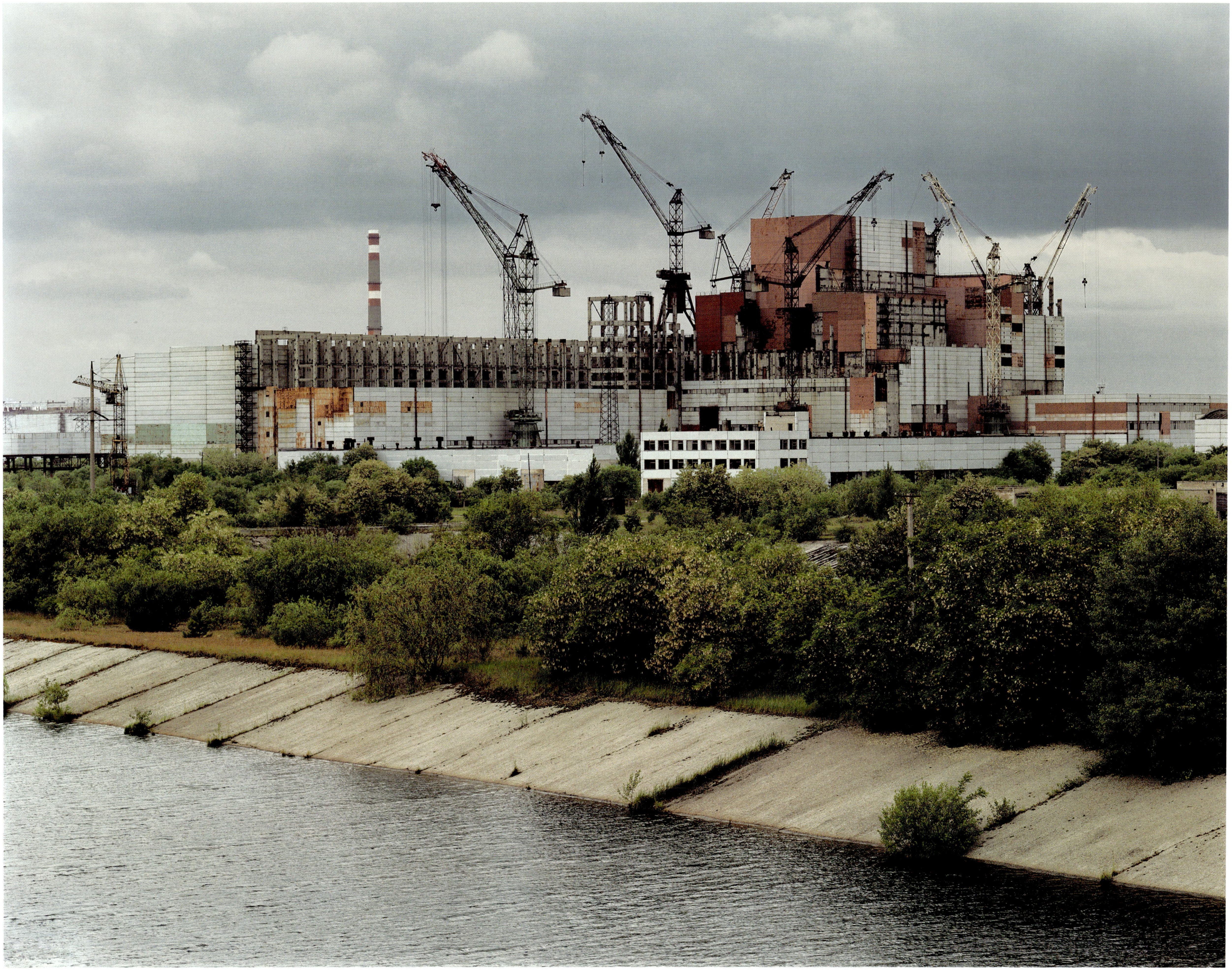

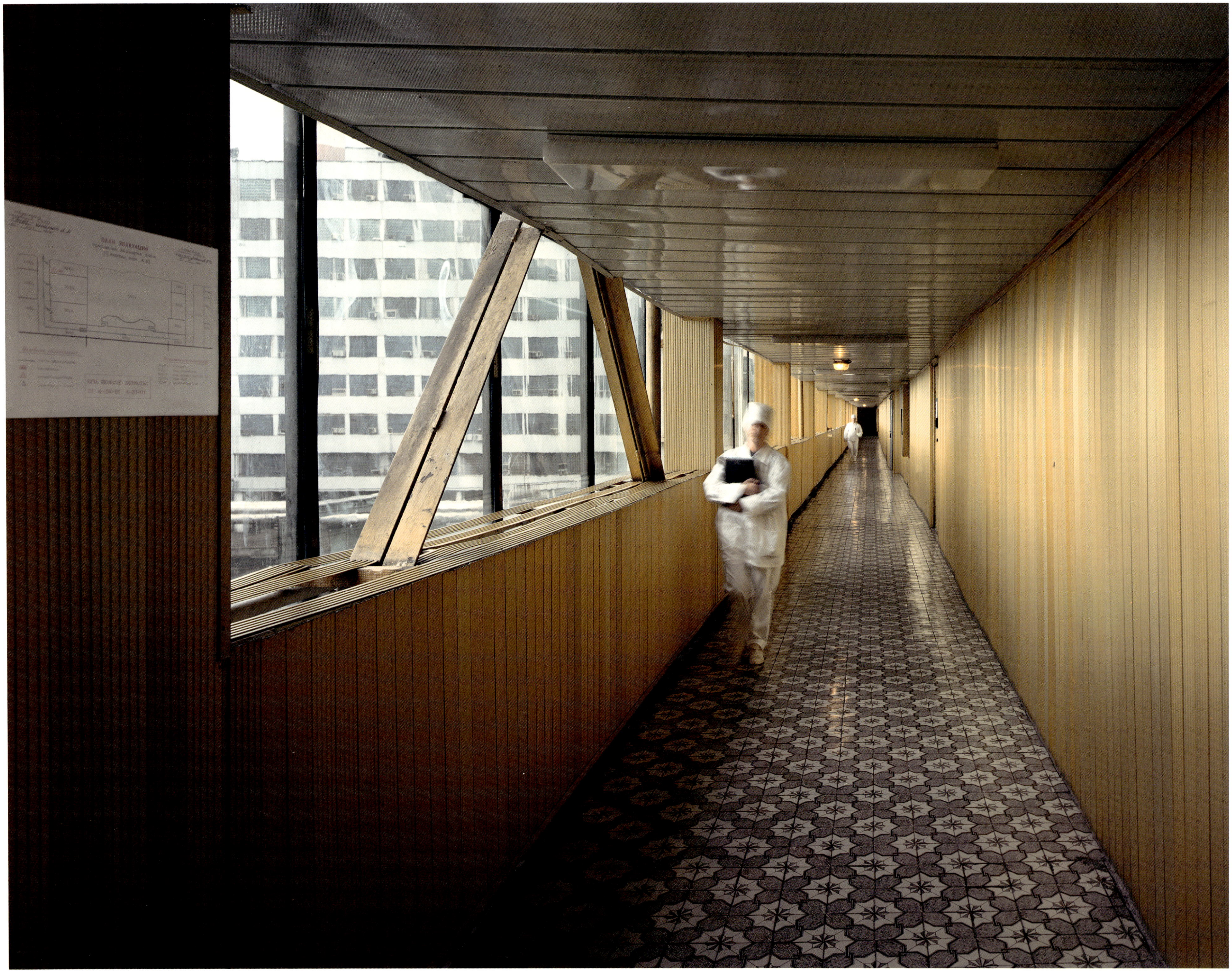
ПЛАН ЭВАКУАЦИИ
01 4-24-01 4-31-01

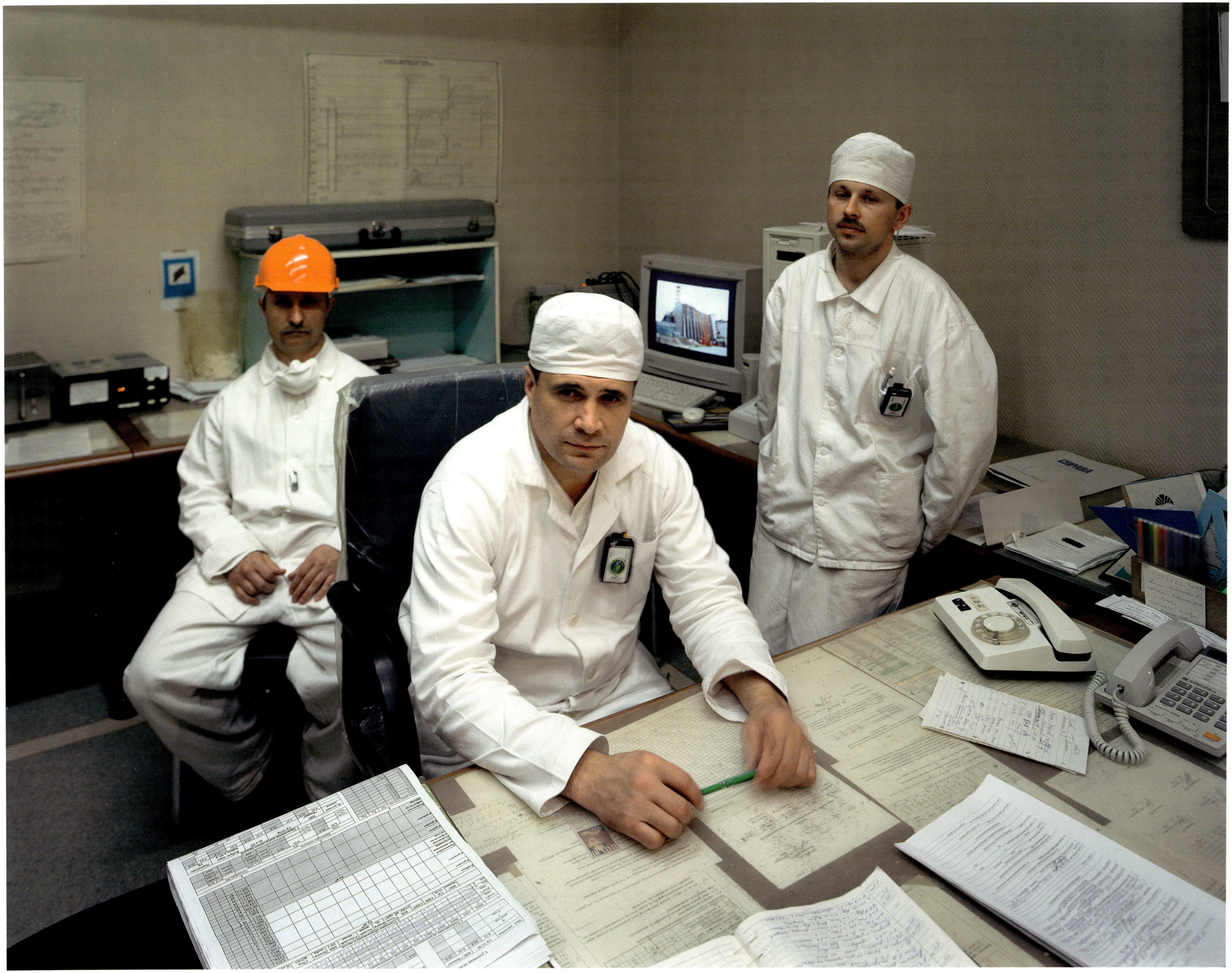

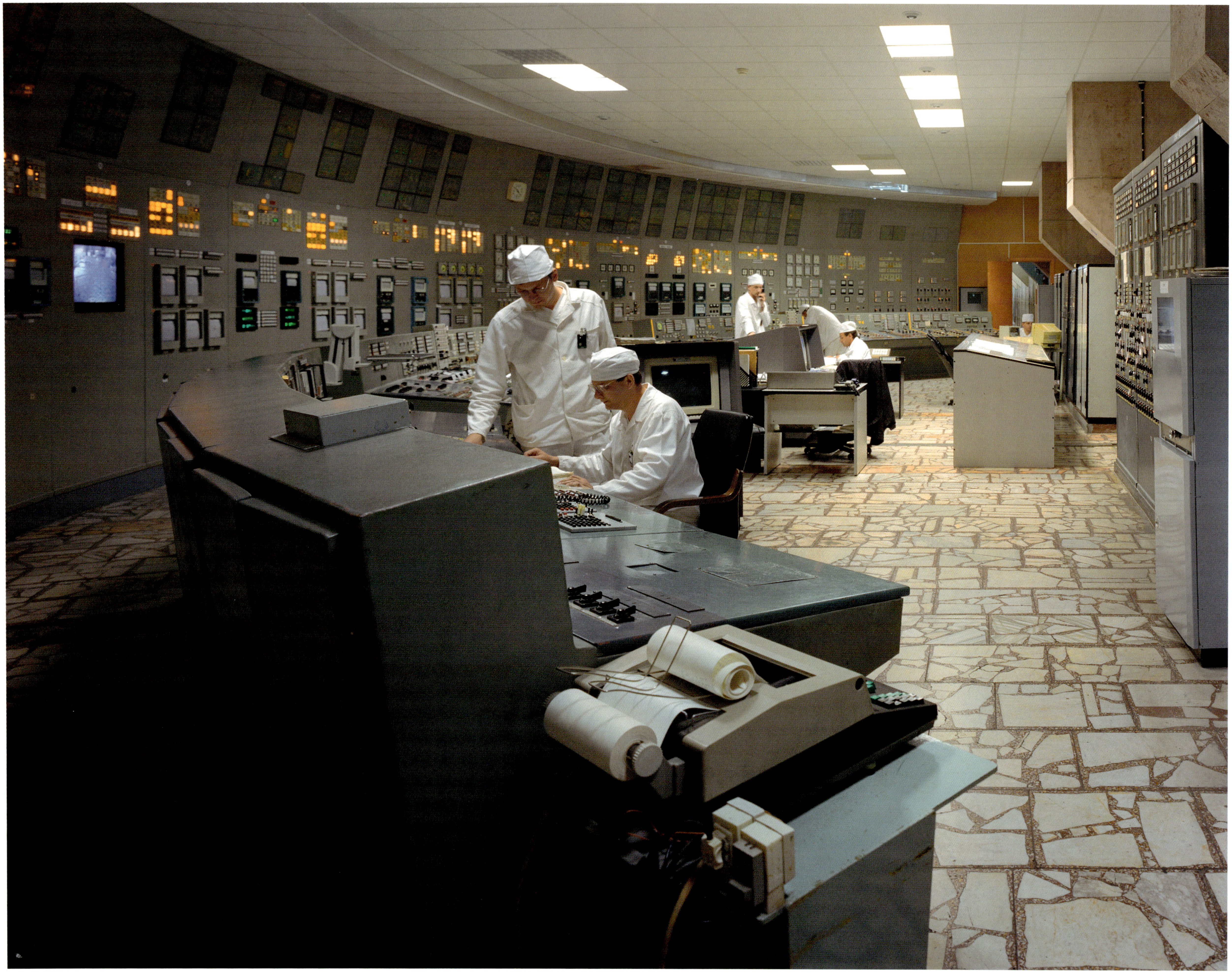

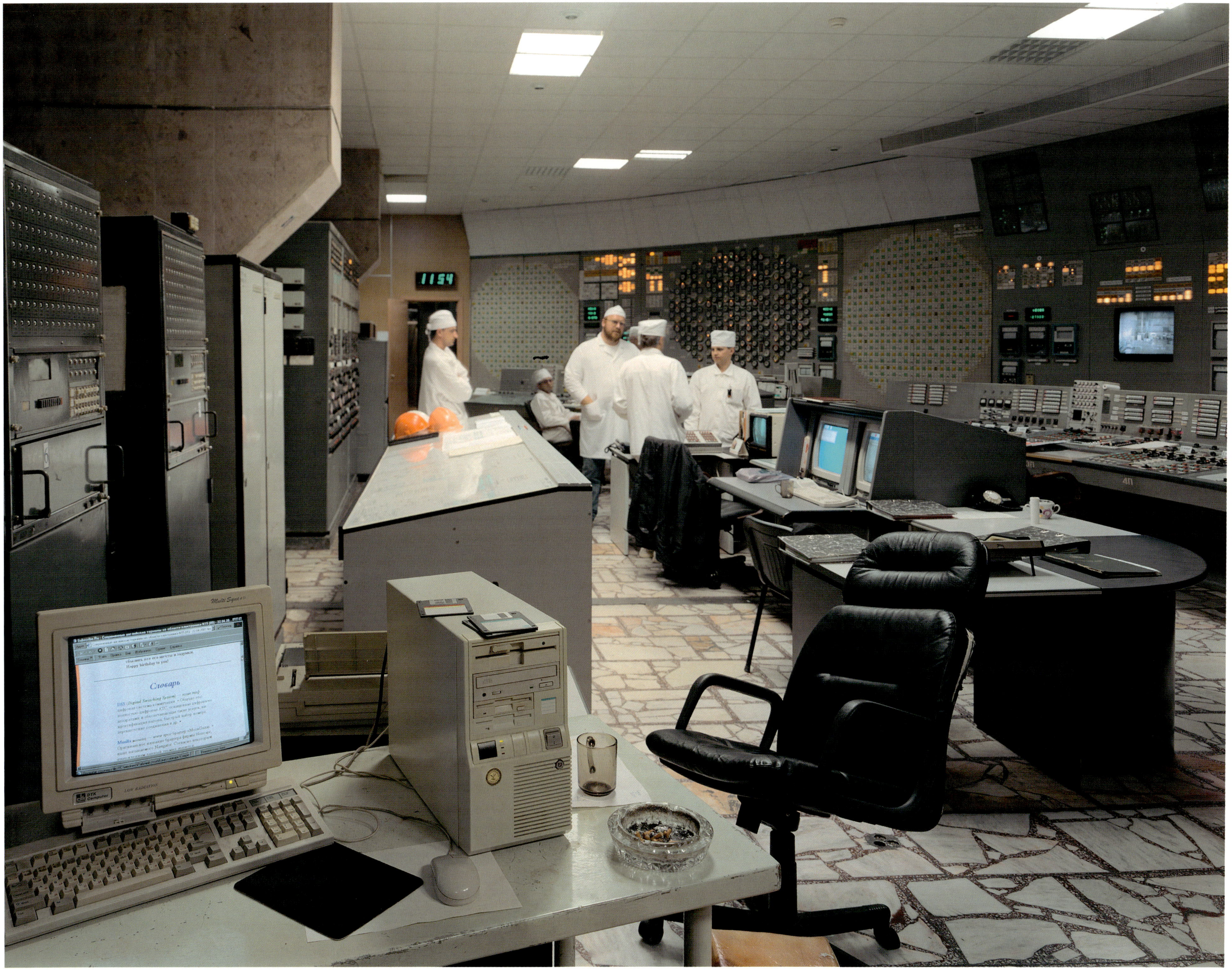
1154
Словарь

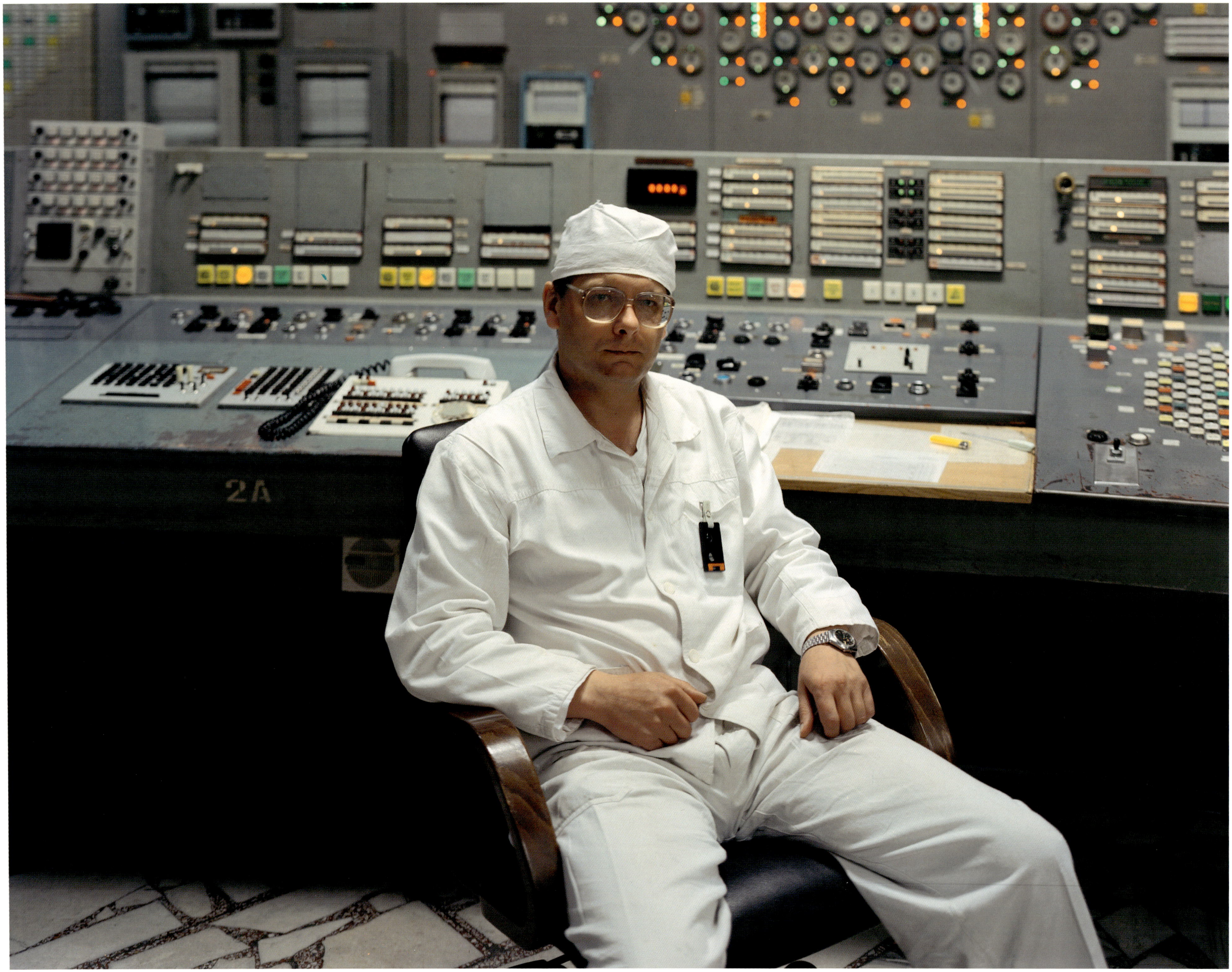
2A

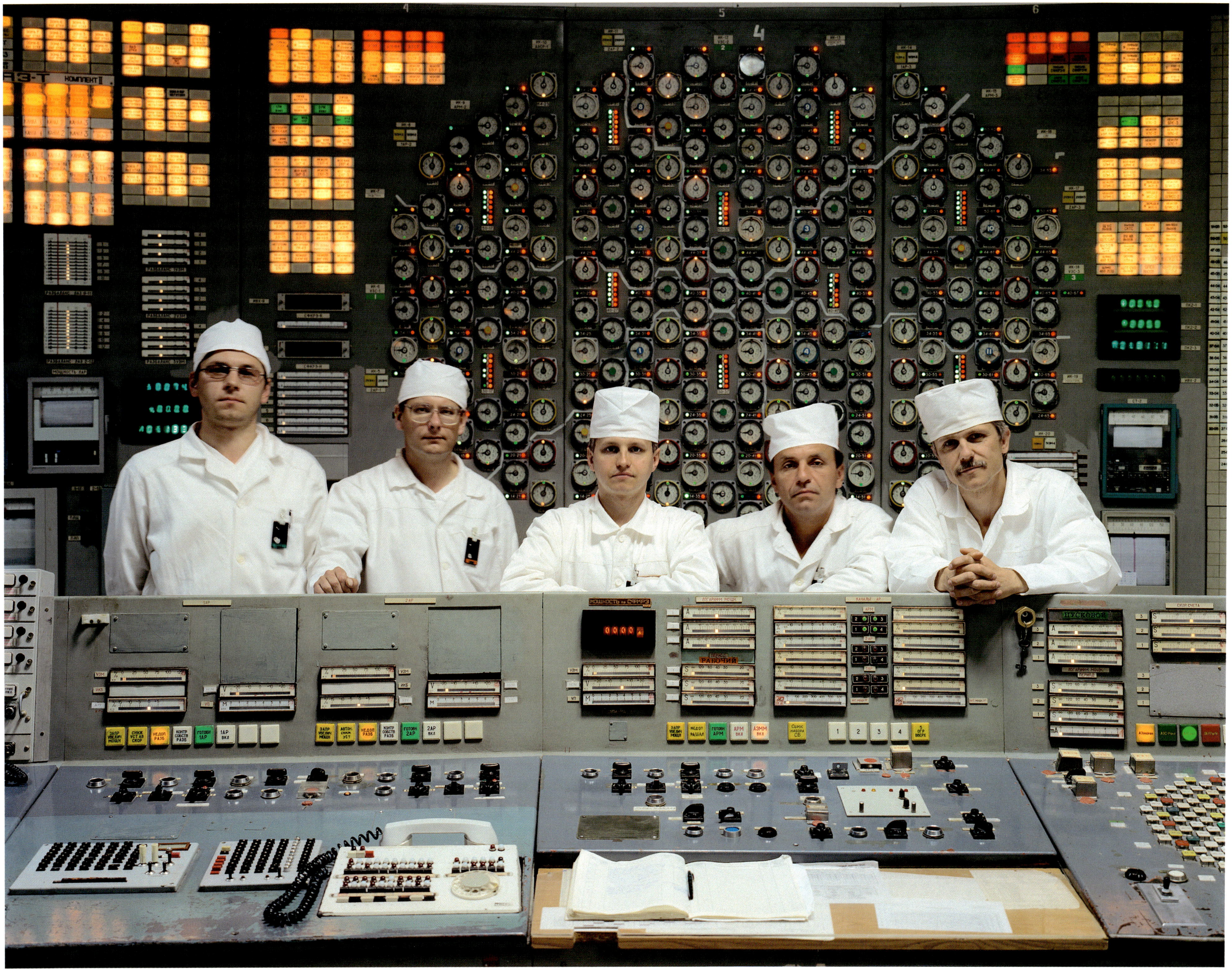

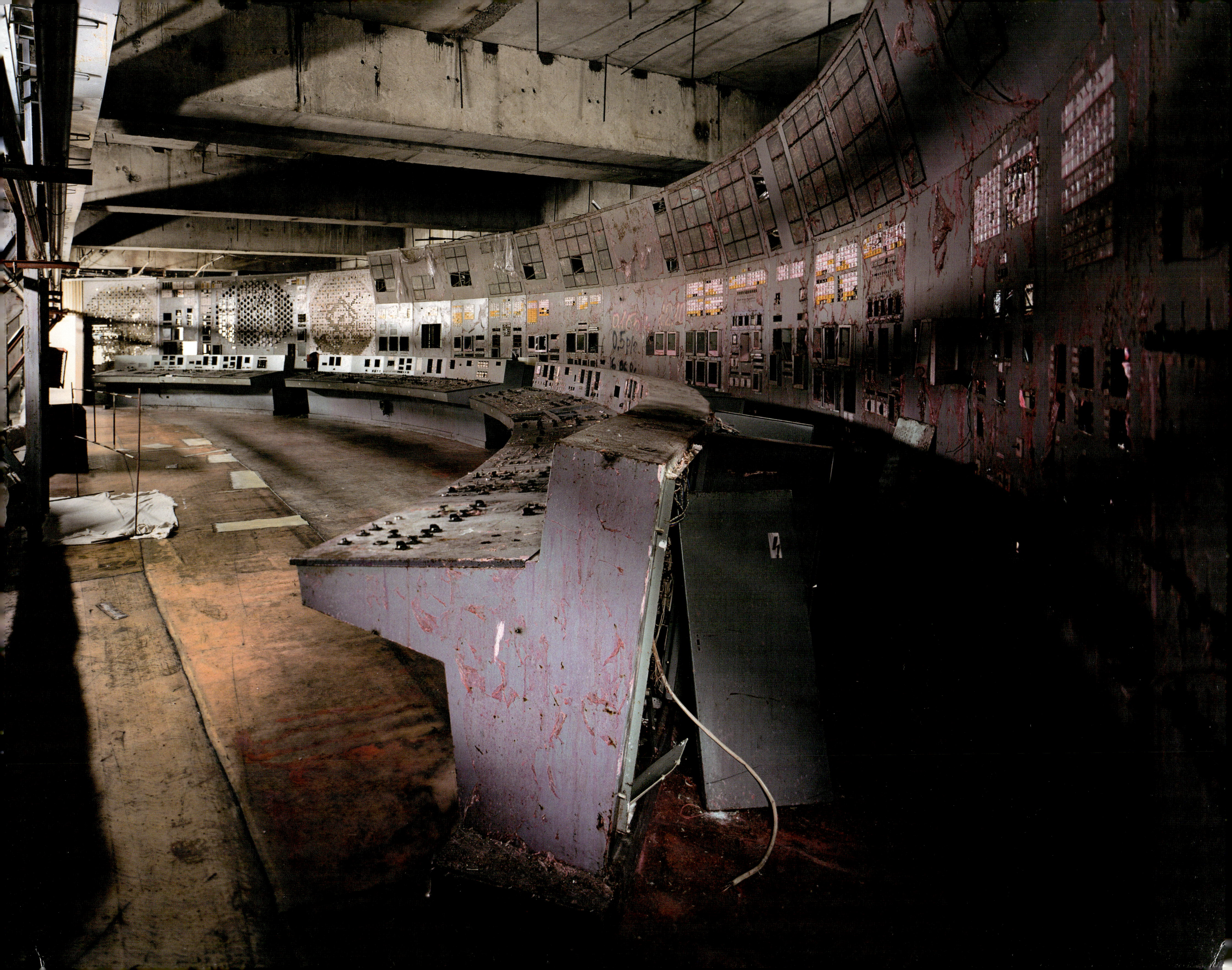

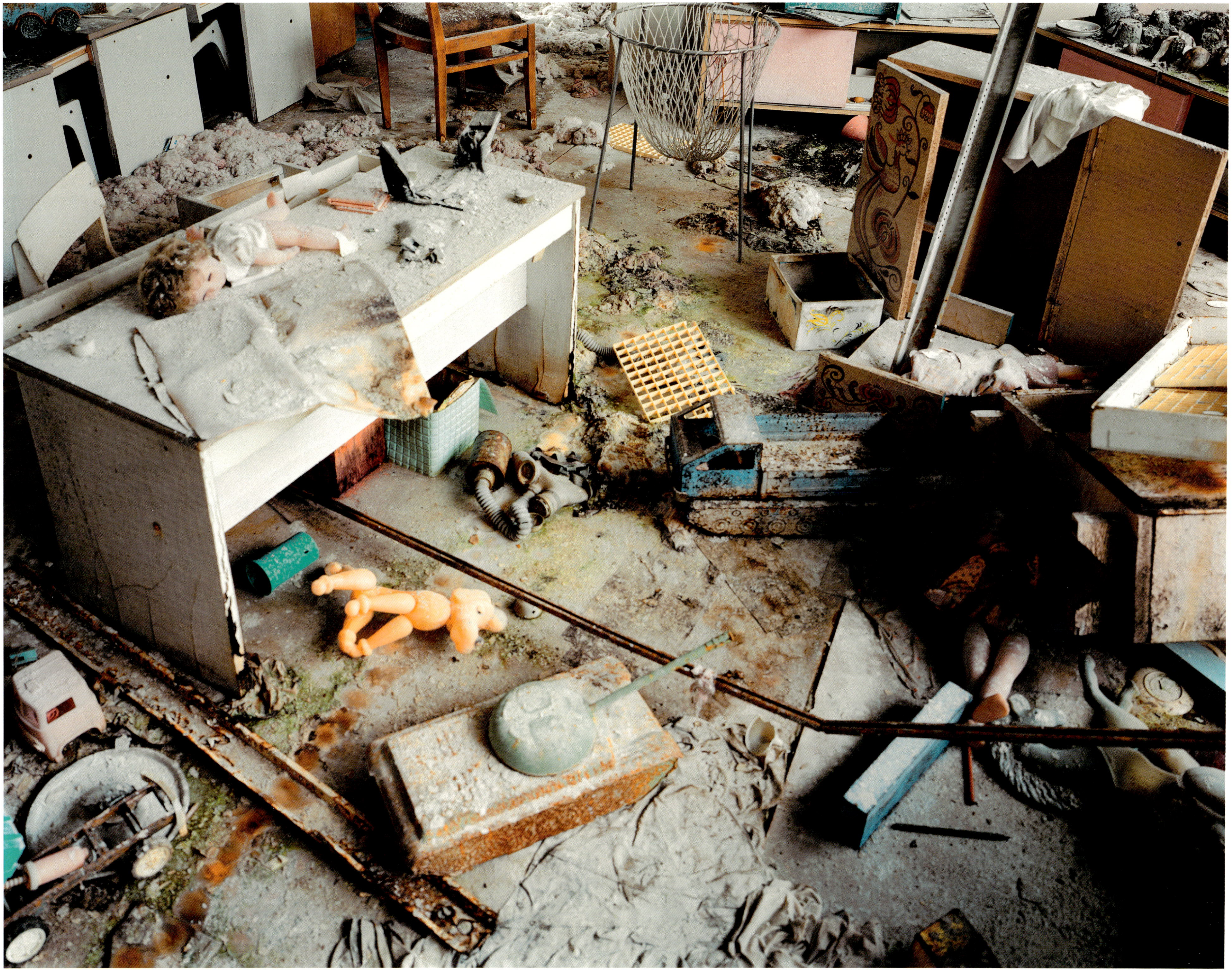

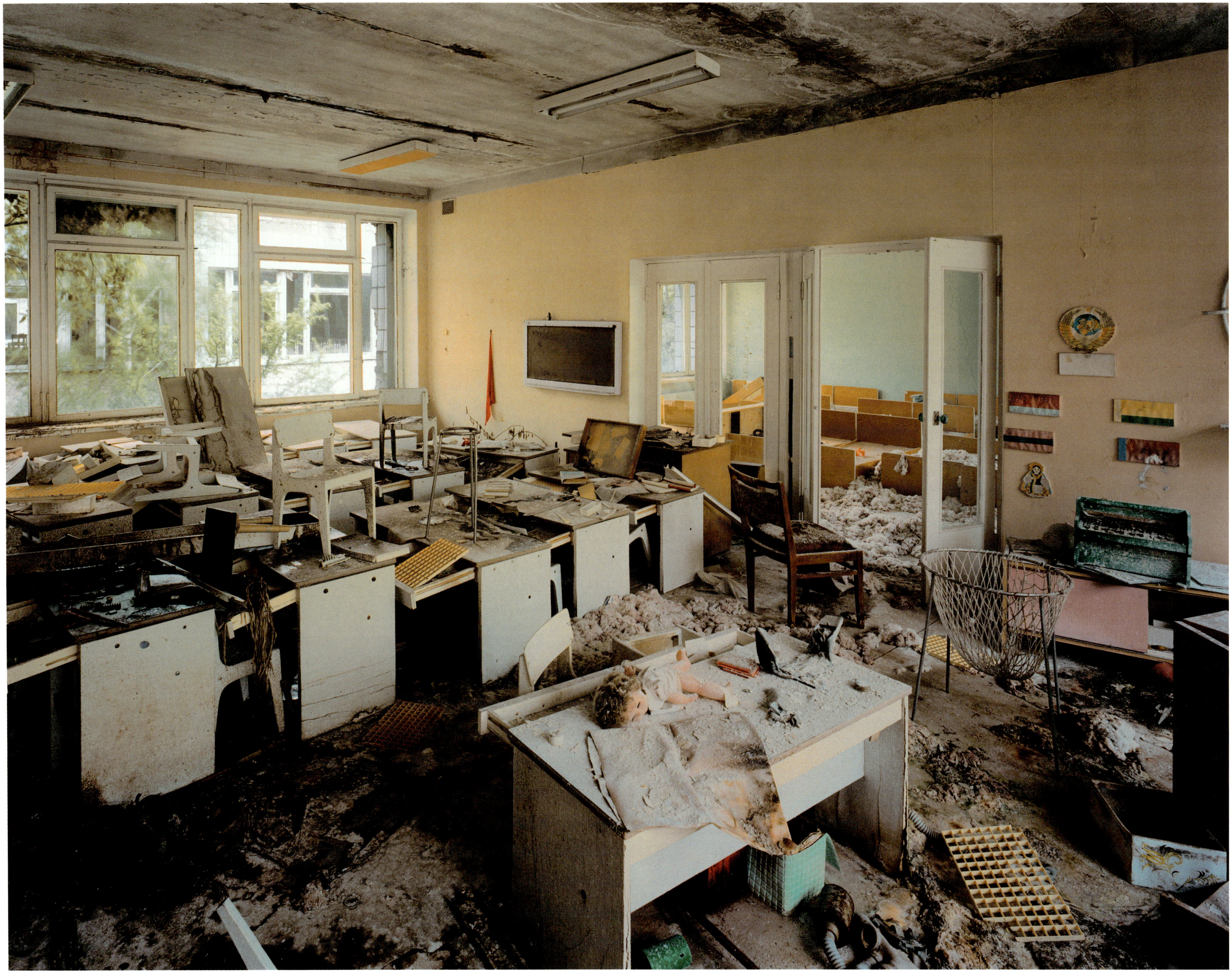

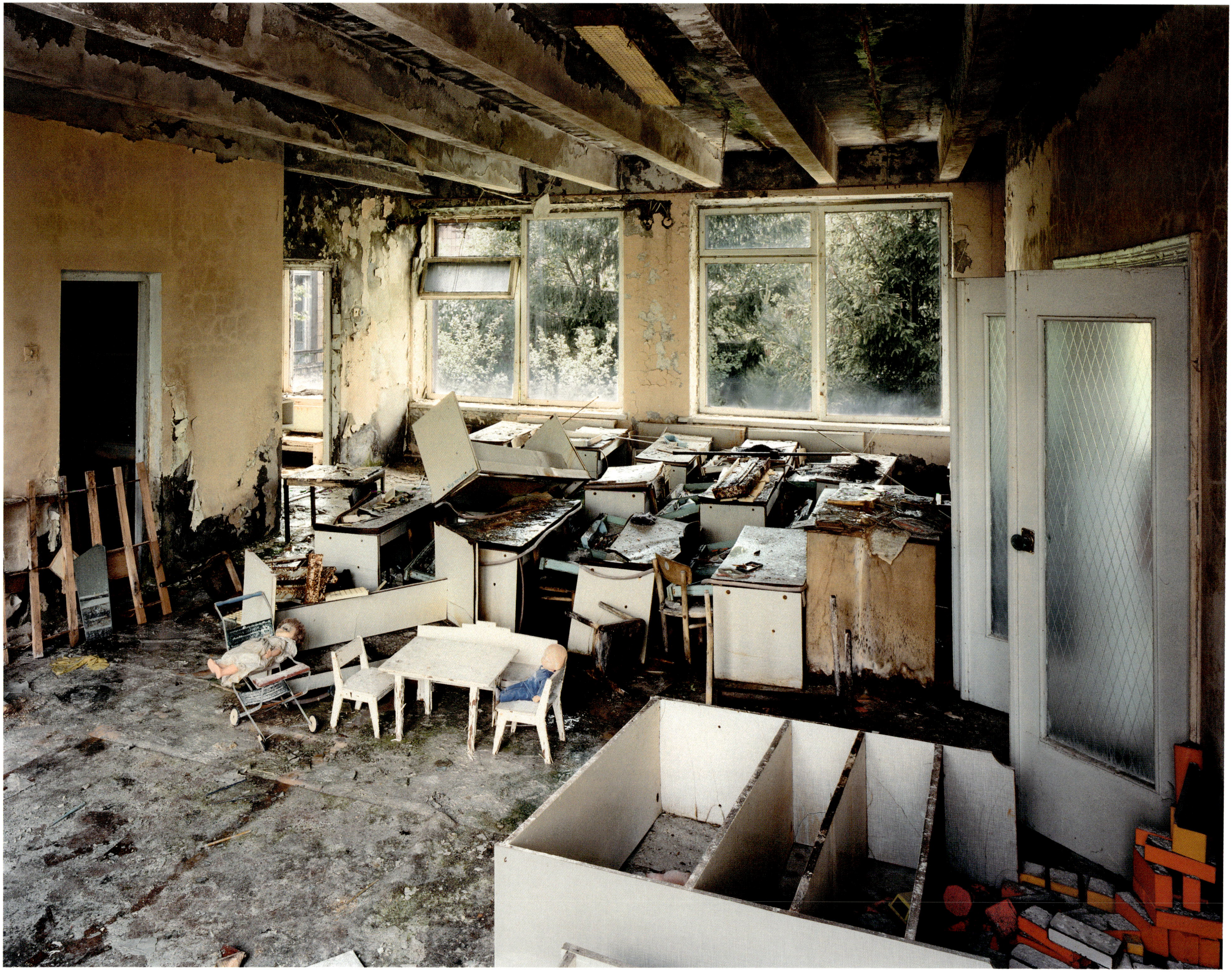

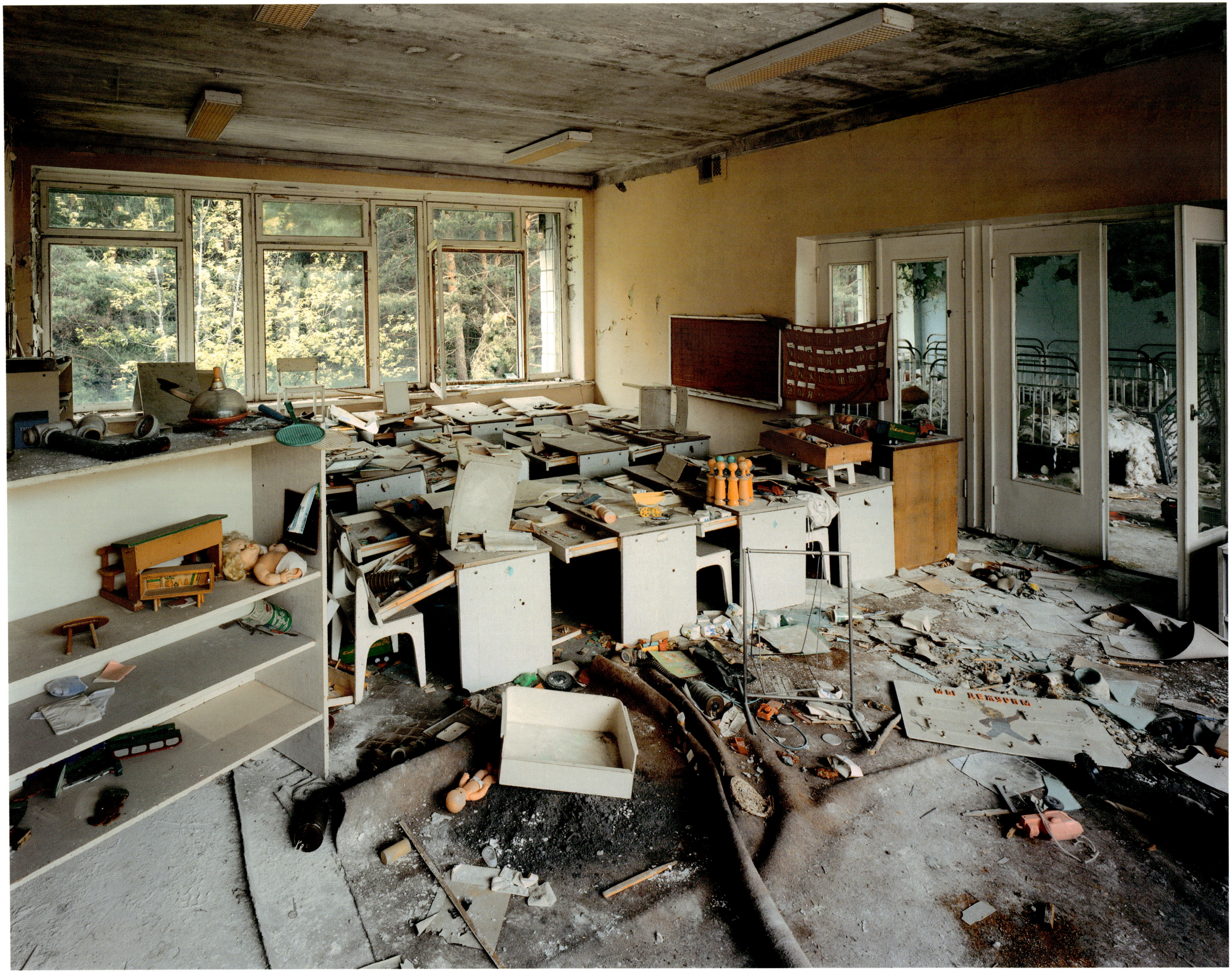
МЫ ДЕЖУРИМ

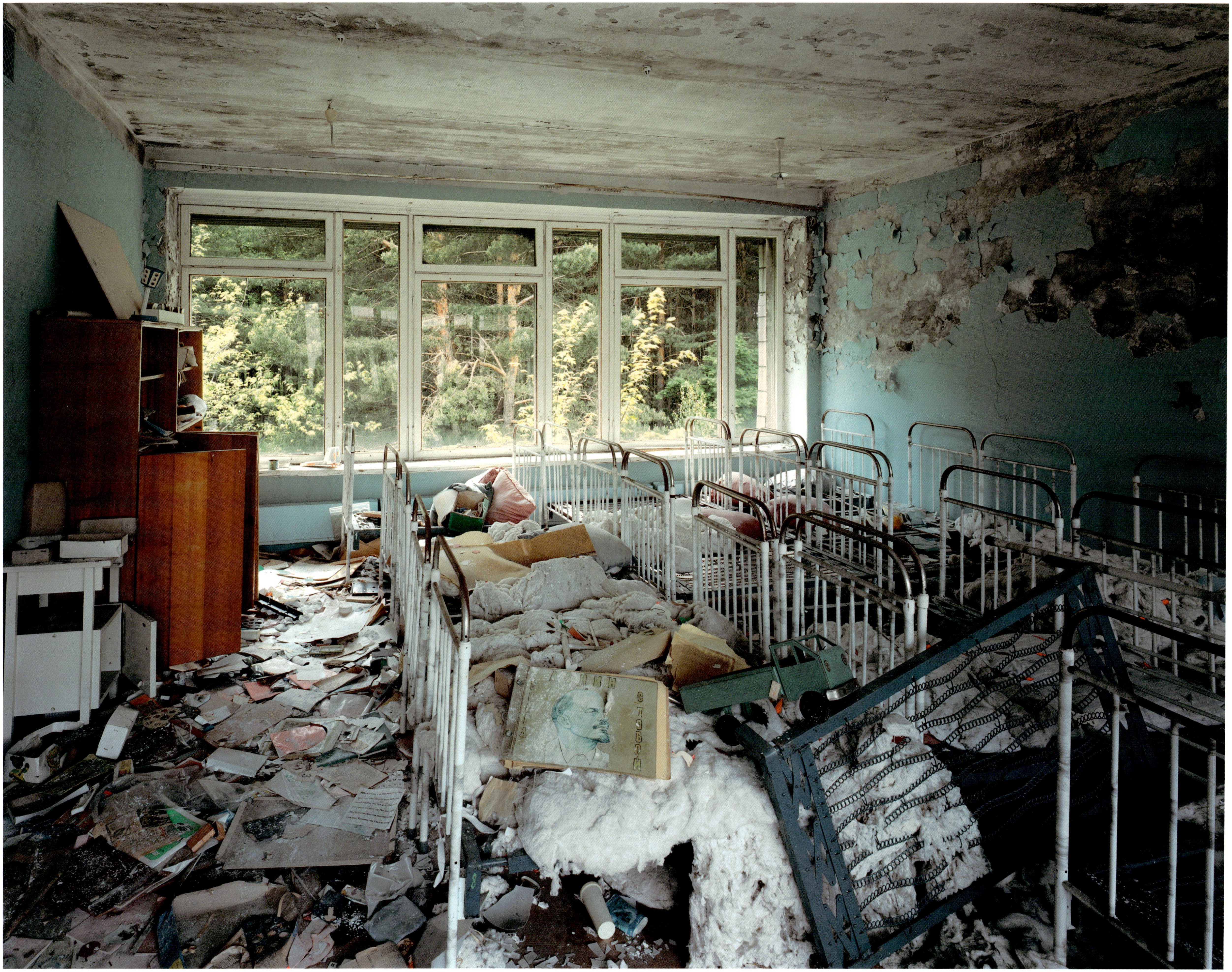

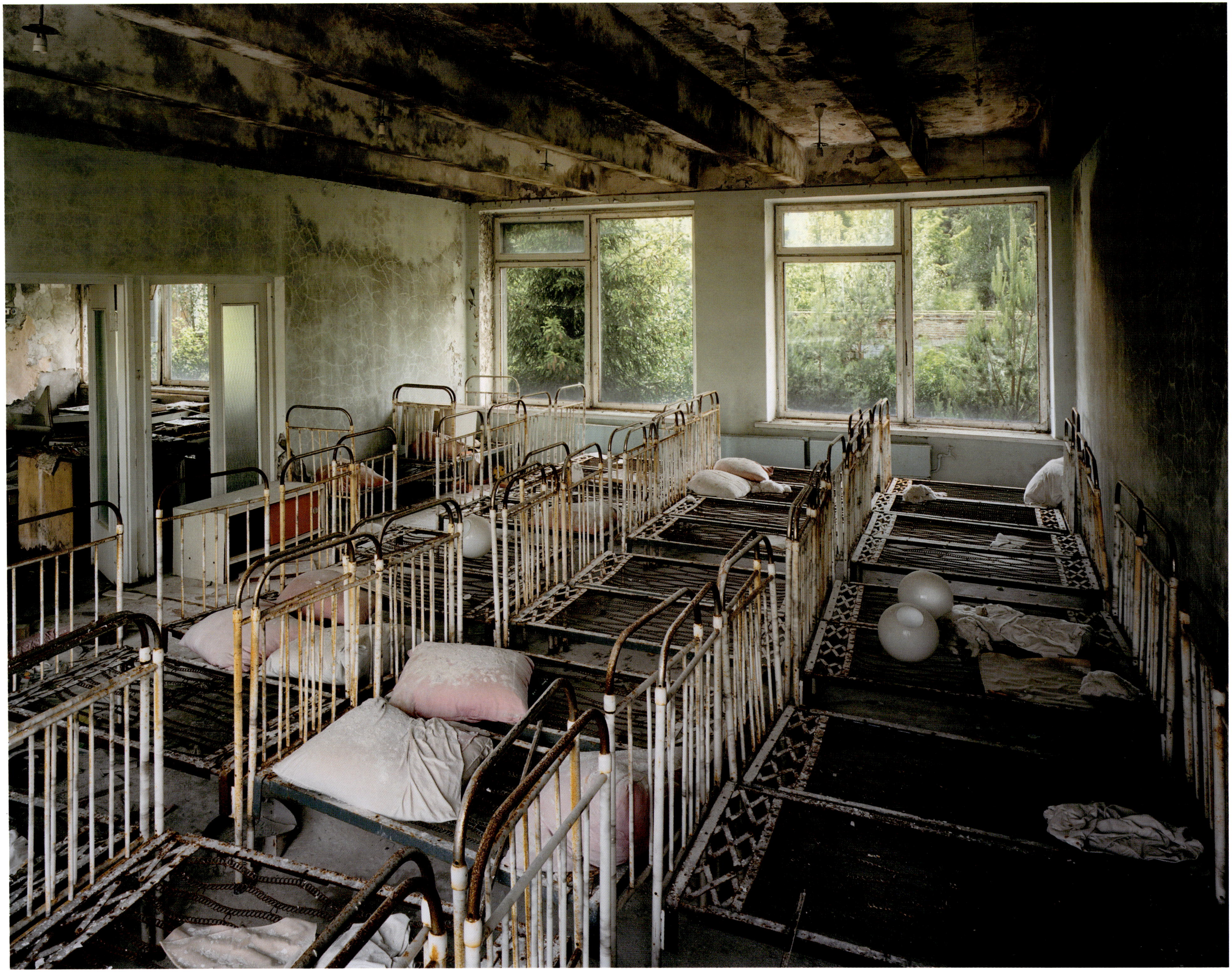

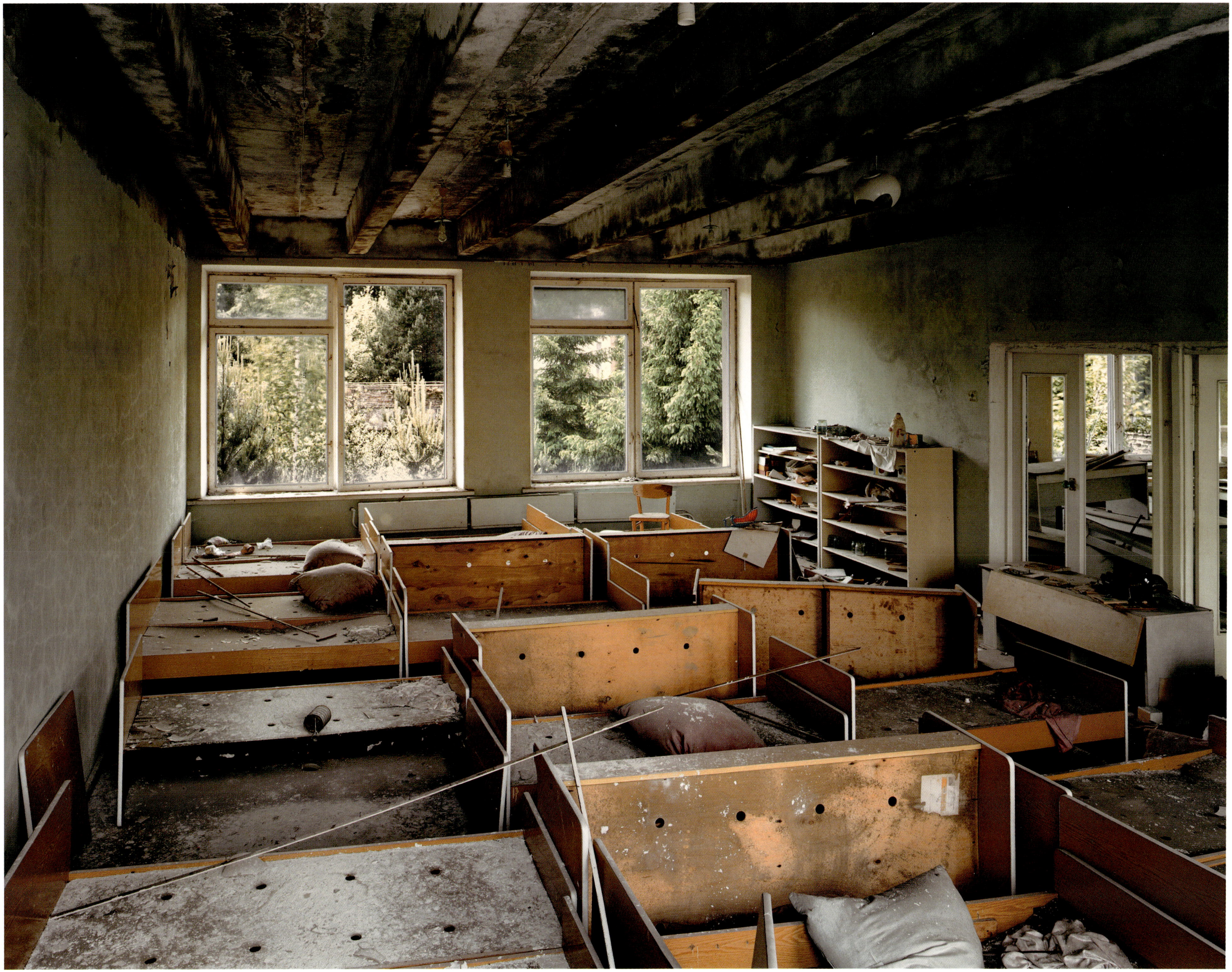

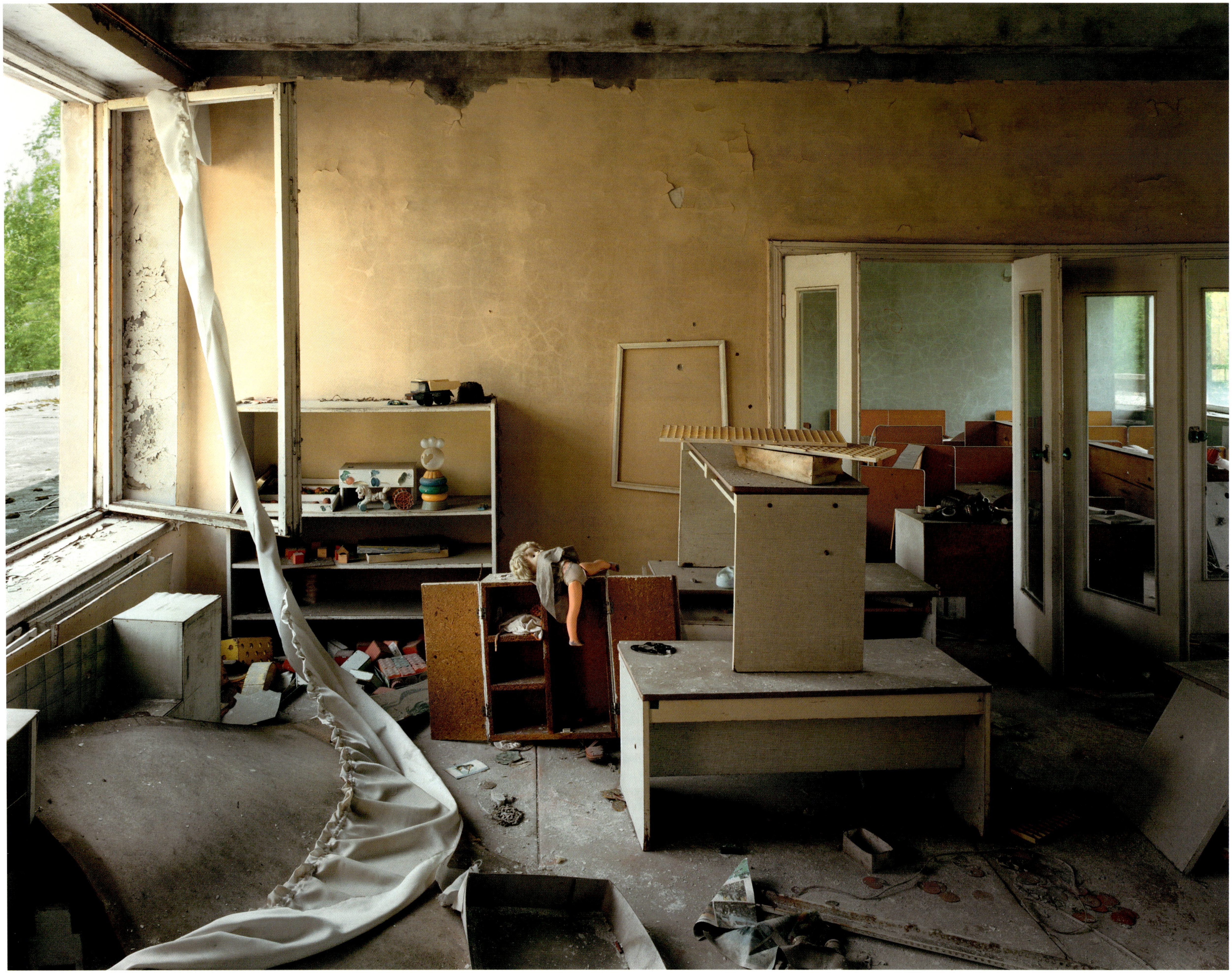

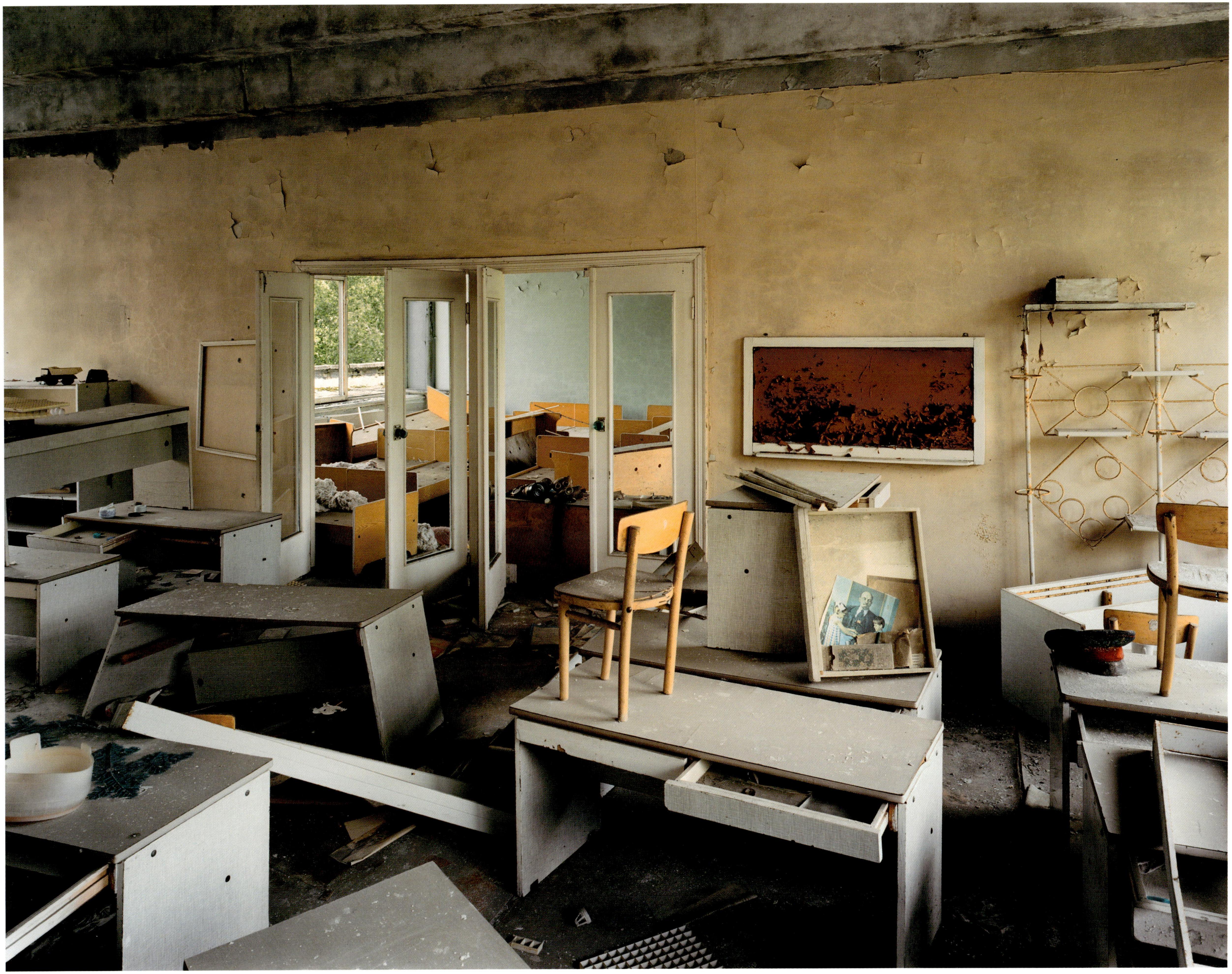

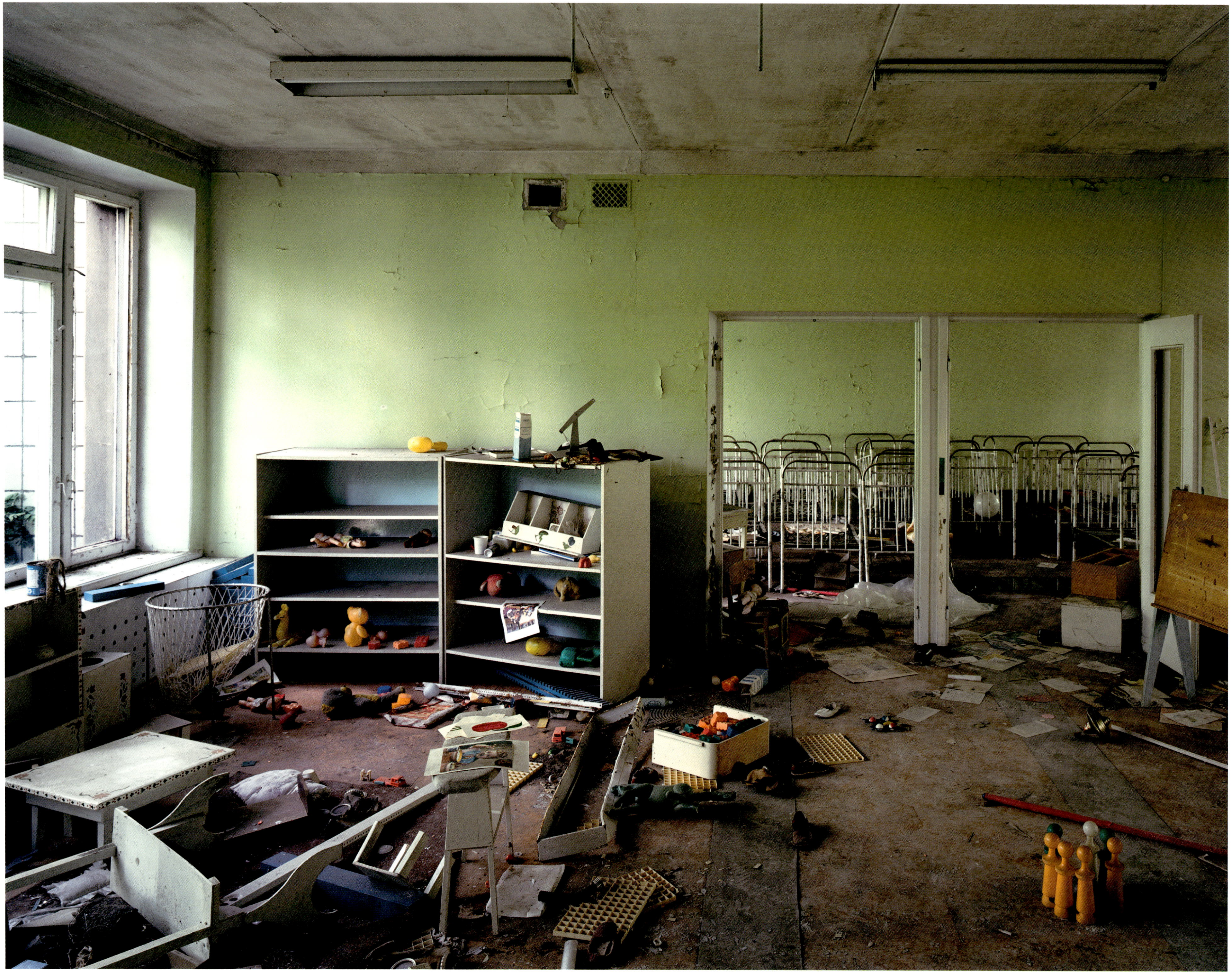

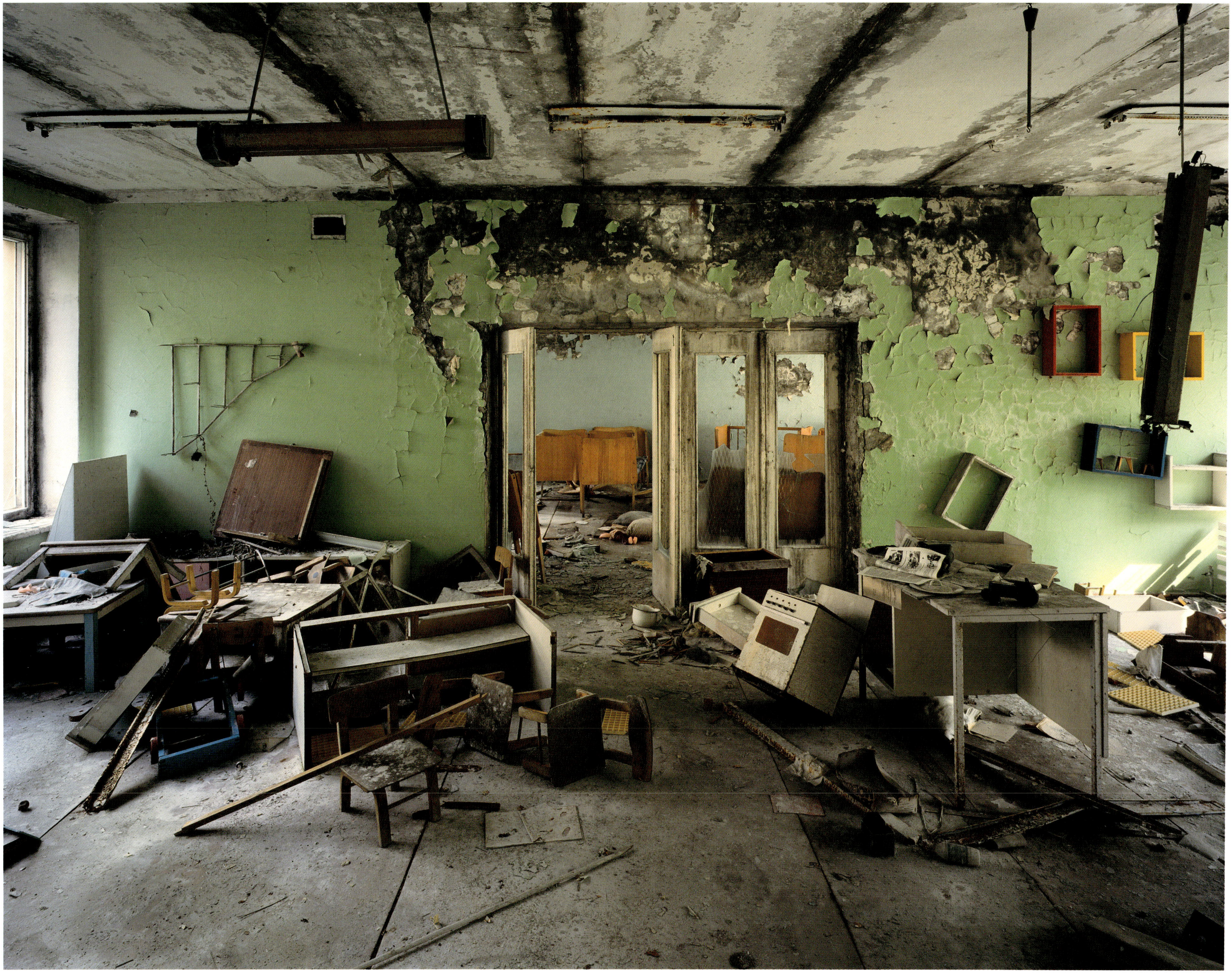

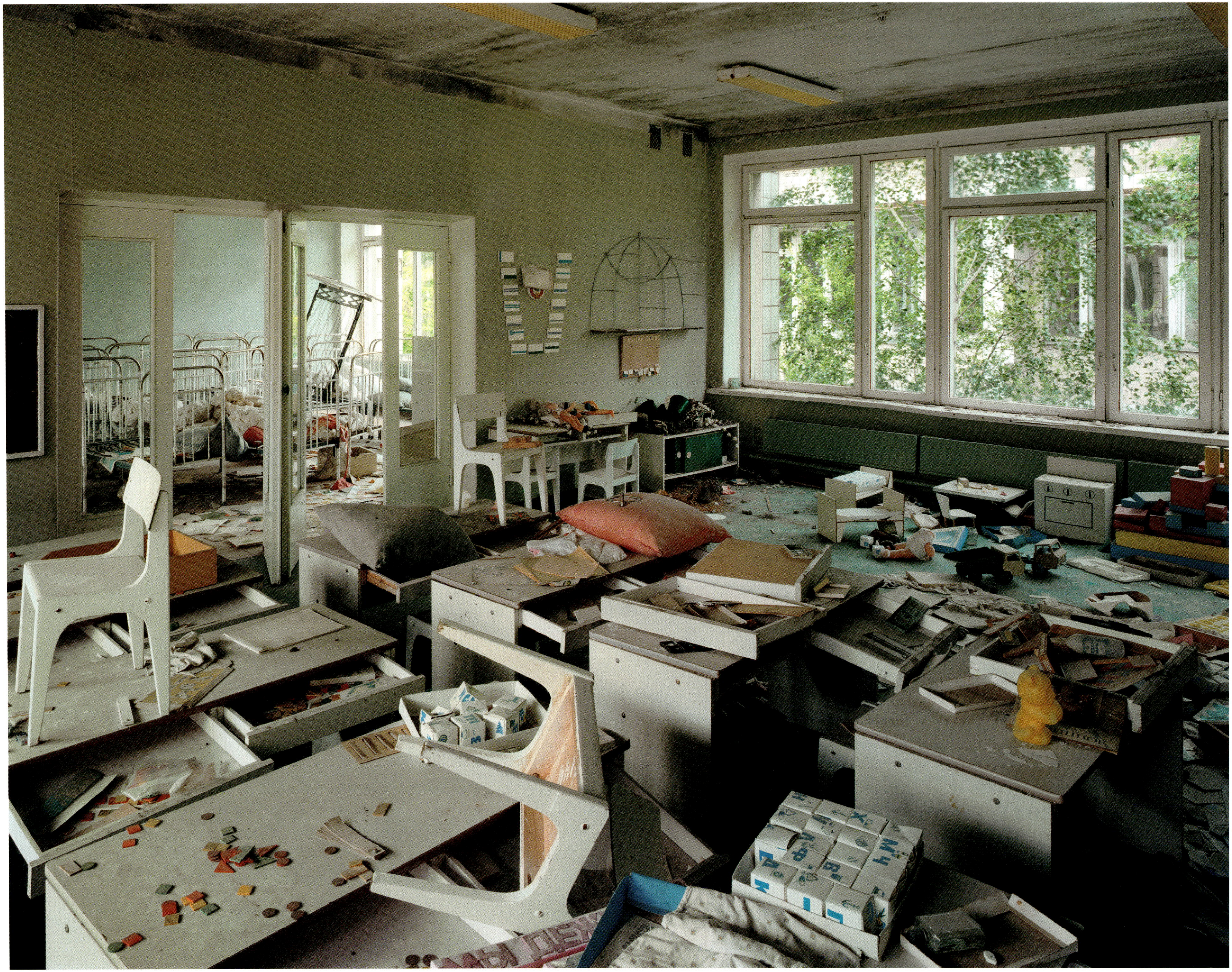

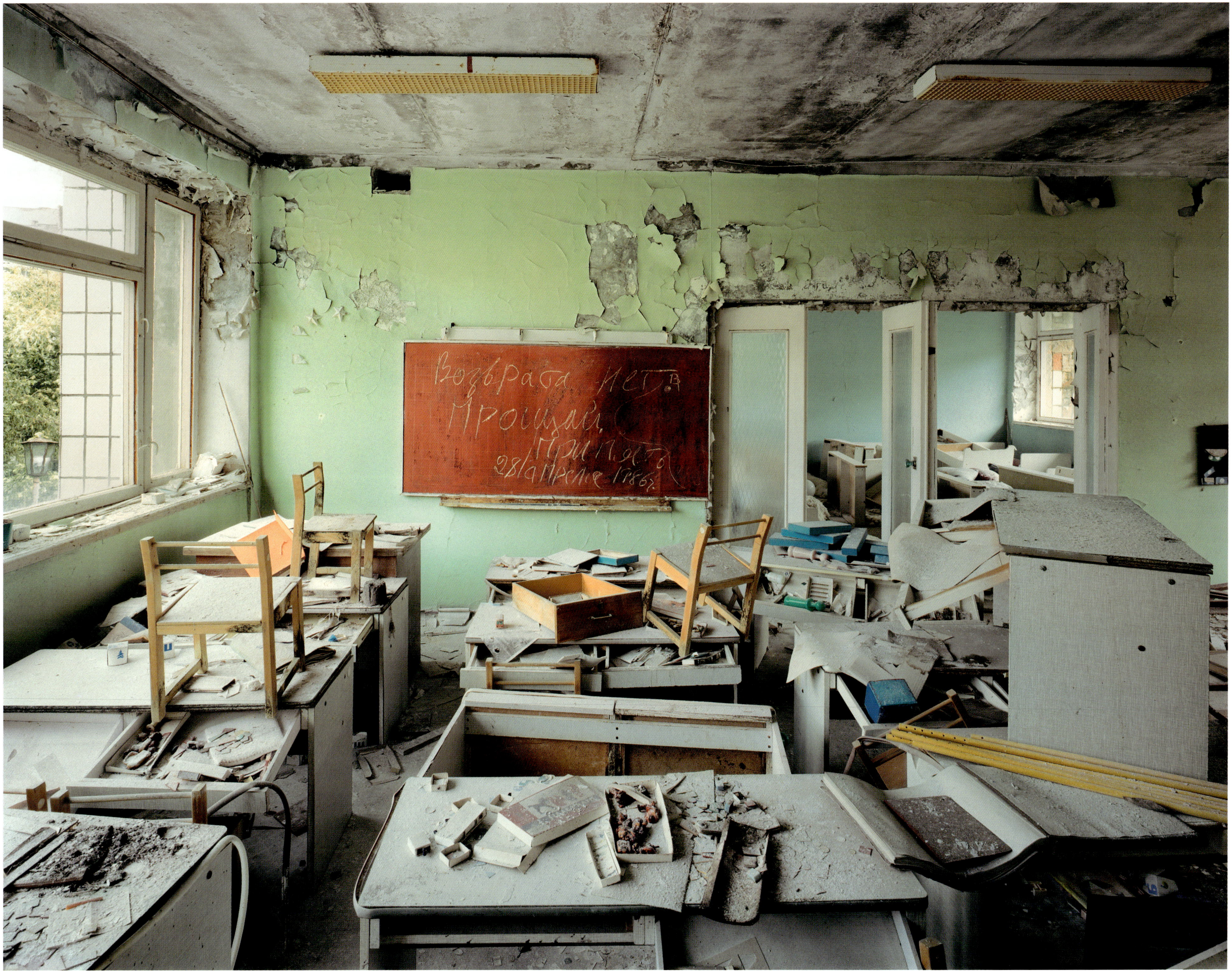
Возврата нет
Прощай,
Припять
28 апреля 1986 г.

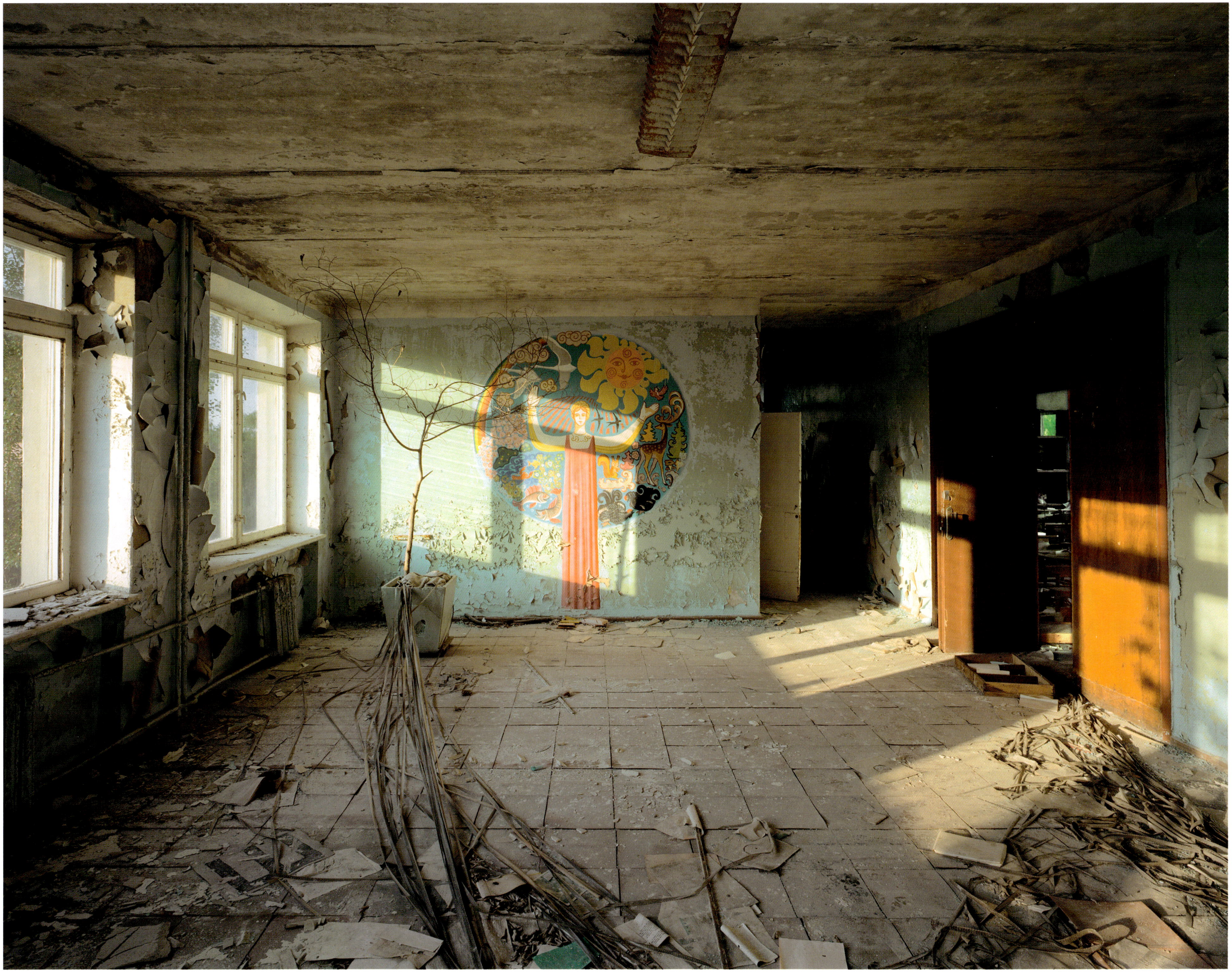

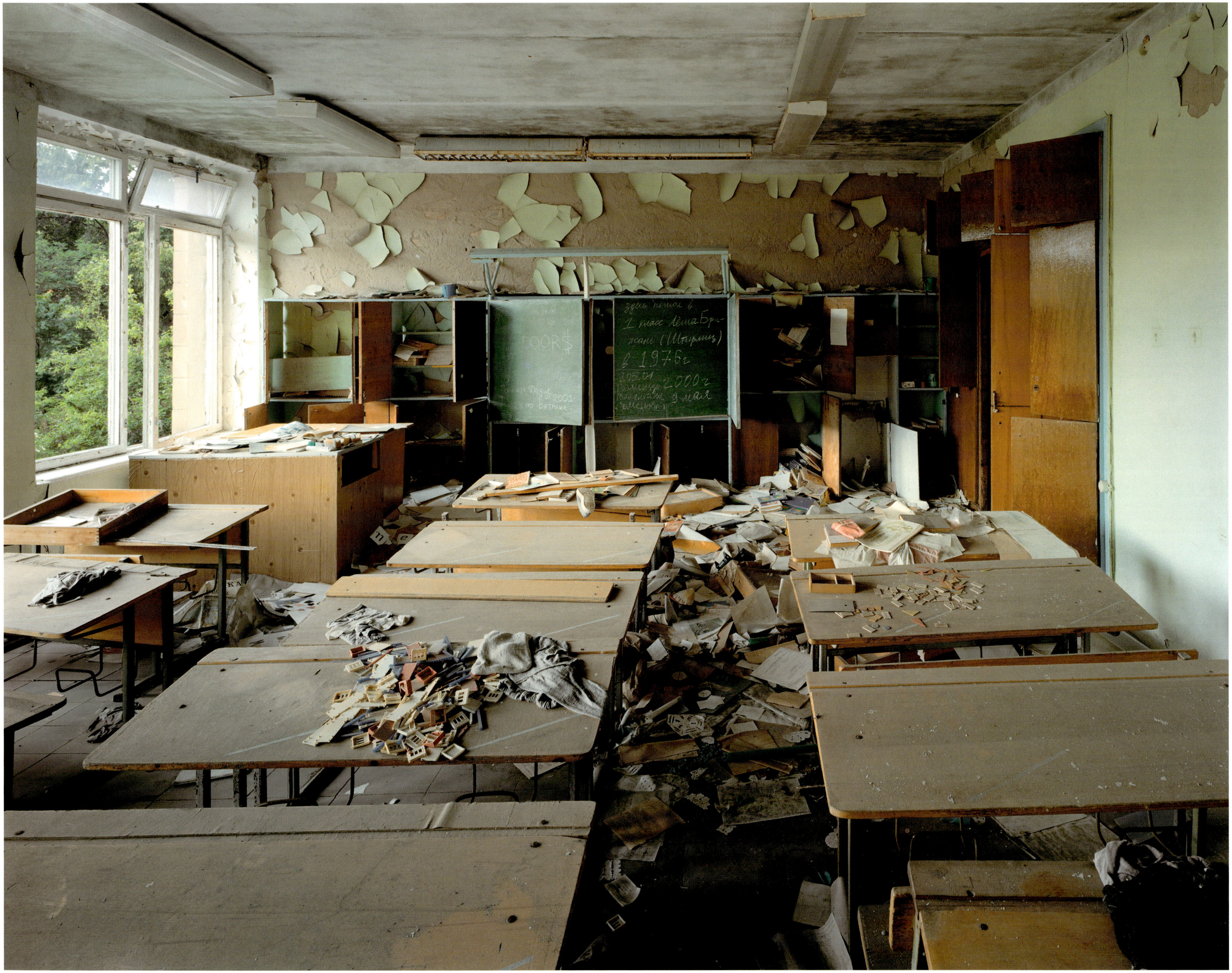
DOOR$
в 1976г
2000г
9 мая

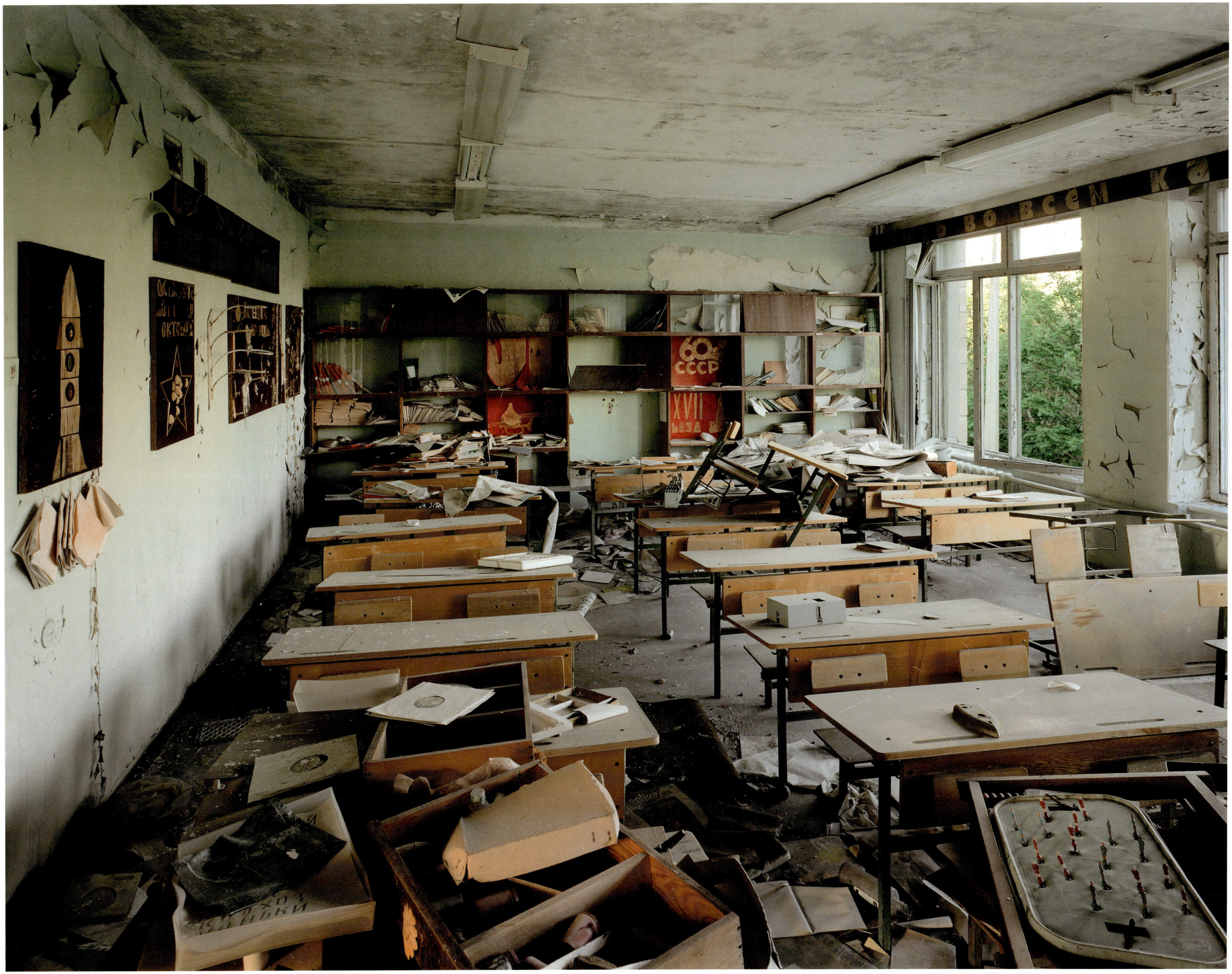
во всем
60
СССР
XVII

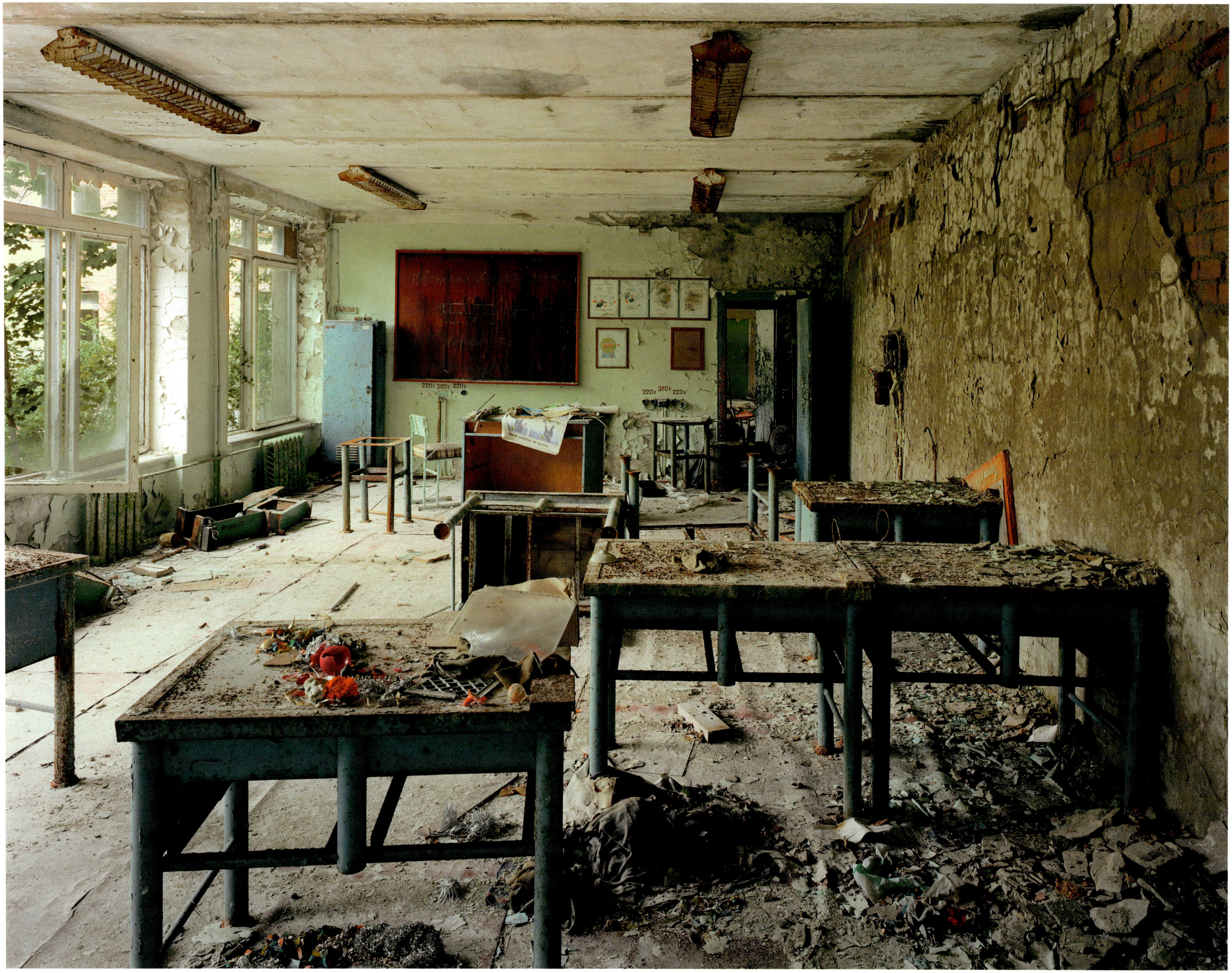
220v 380v 220v
220v 380v 220v

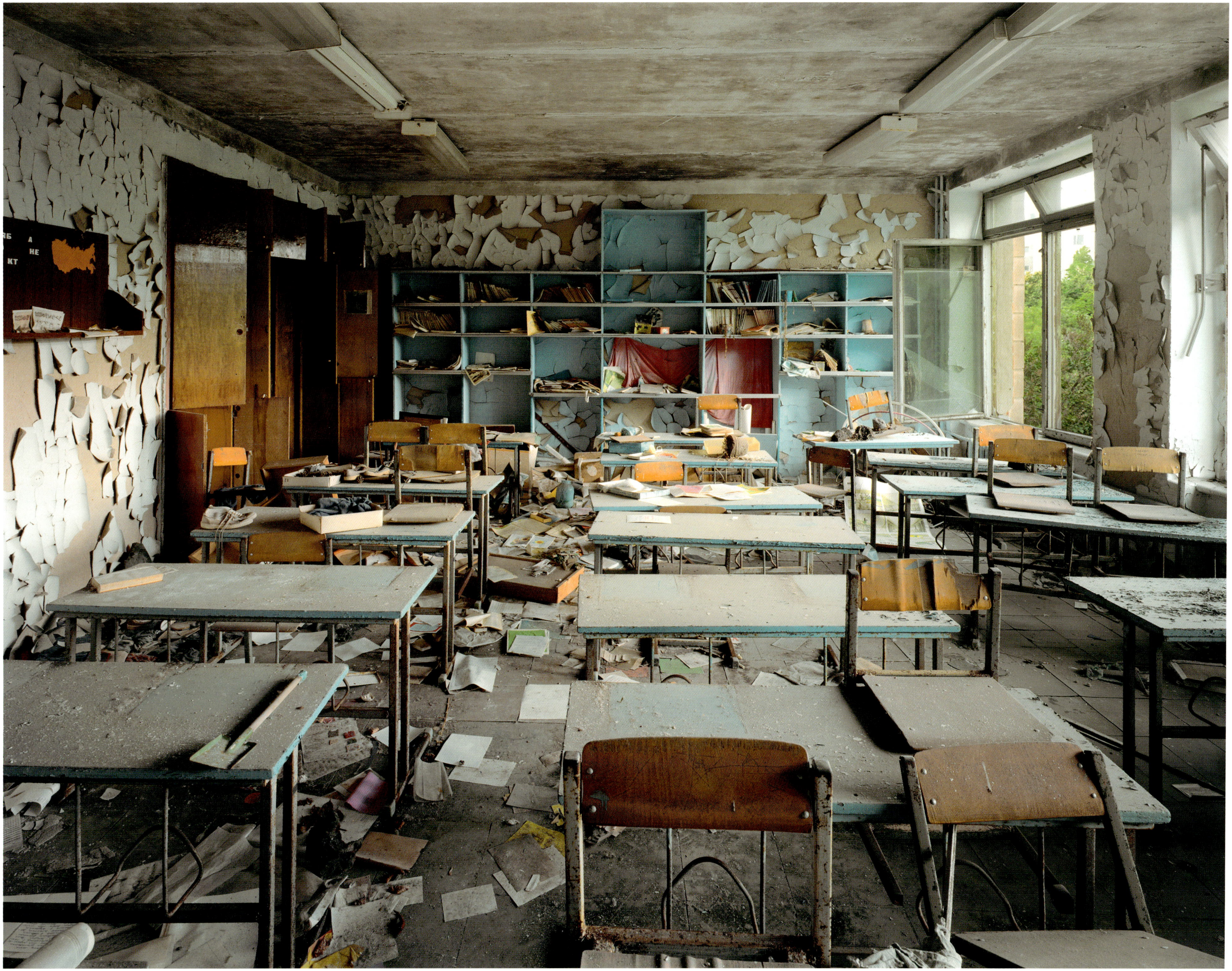

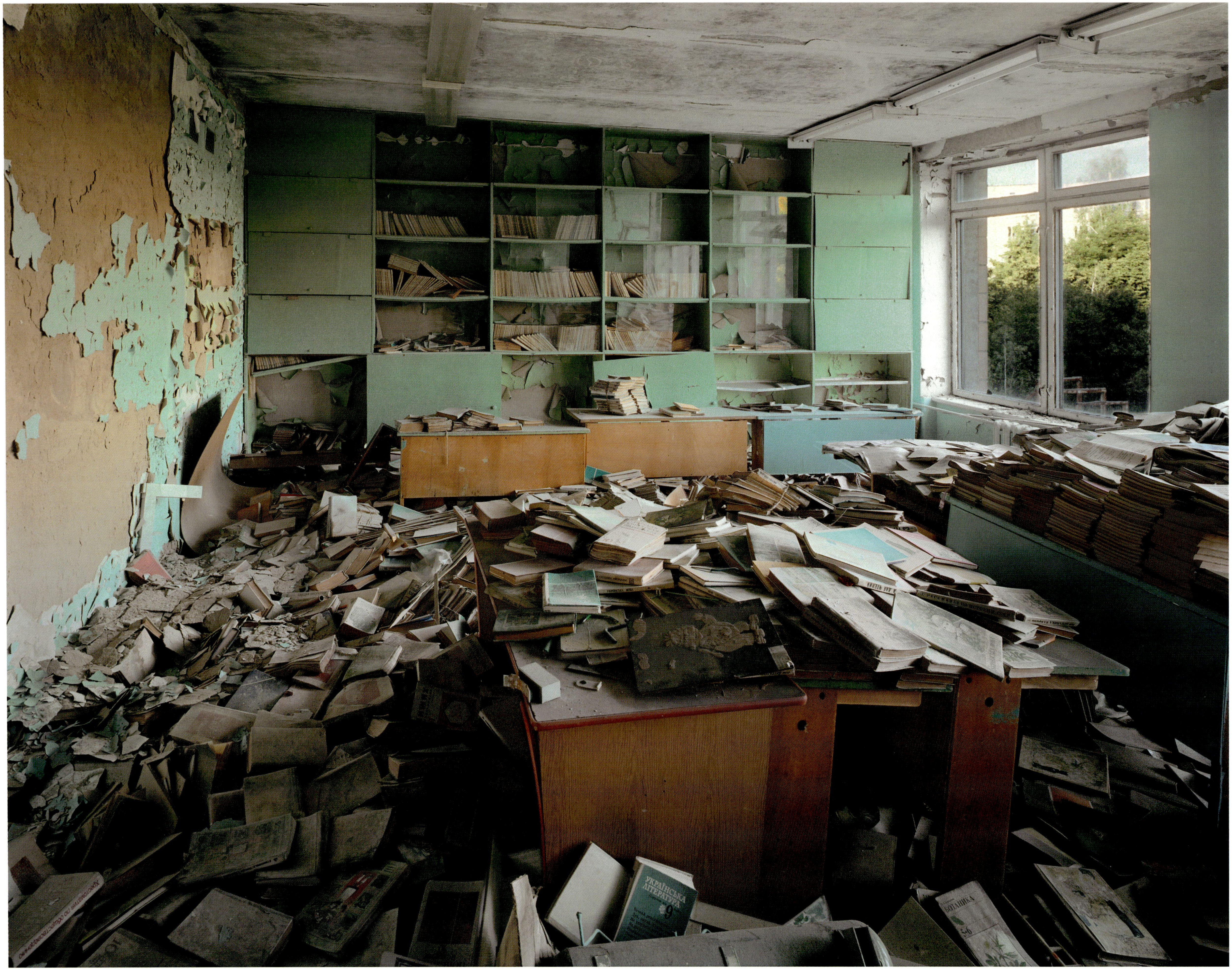
УКРАЇНСЬКА ЛІТЕРАТУРА
9
БОТАНИКА
5-6

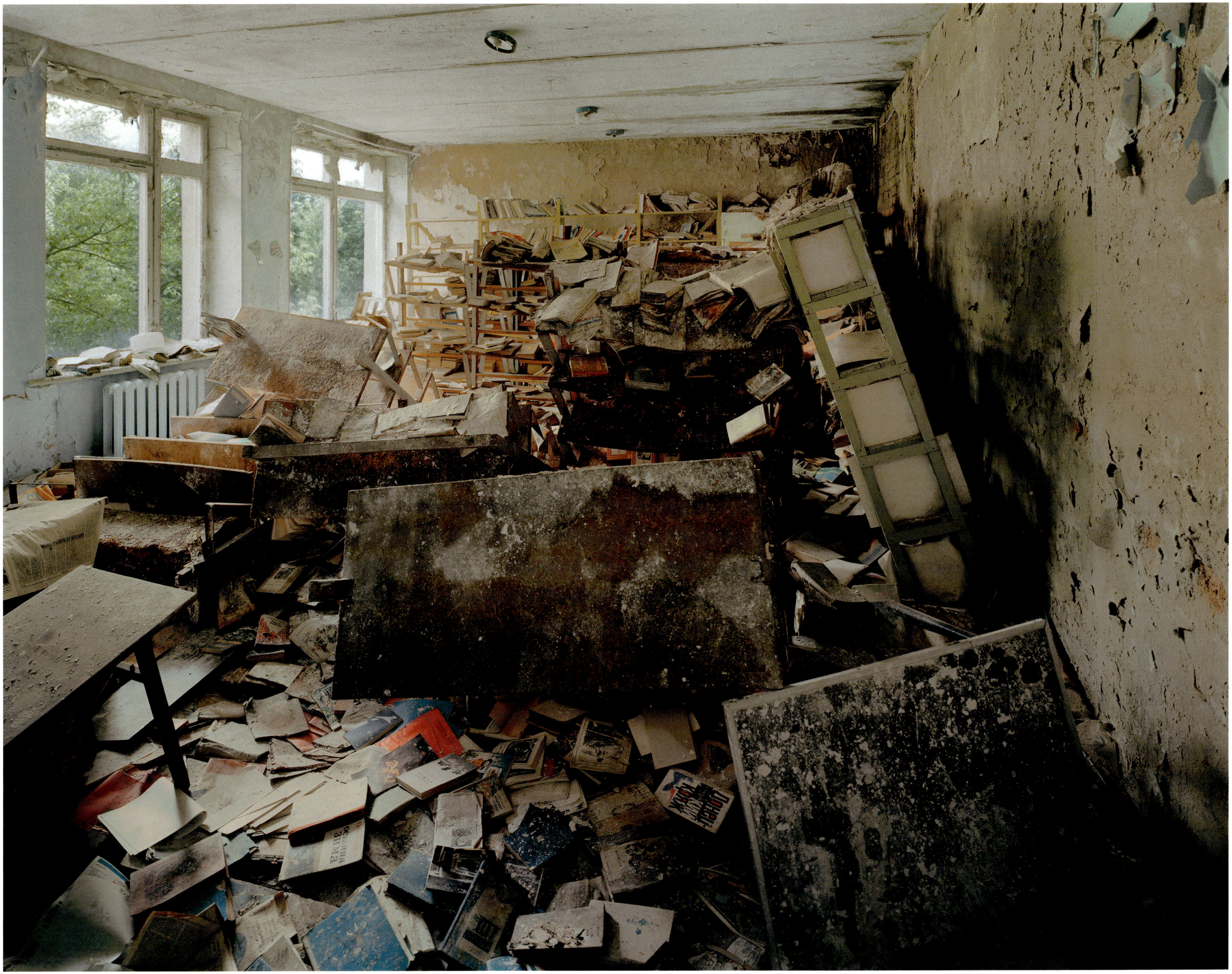

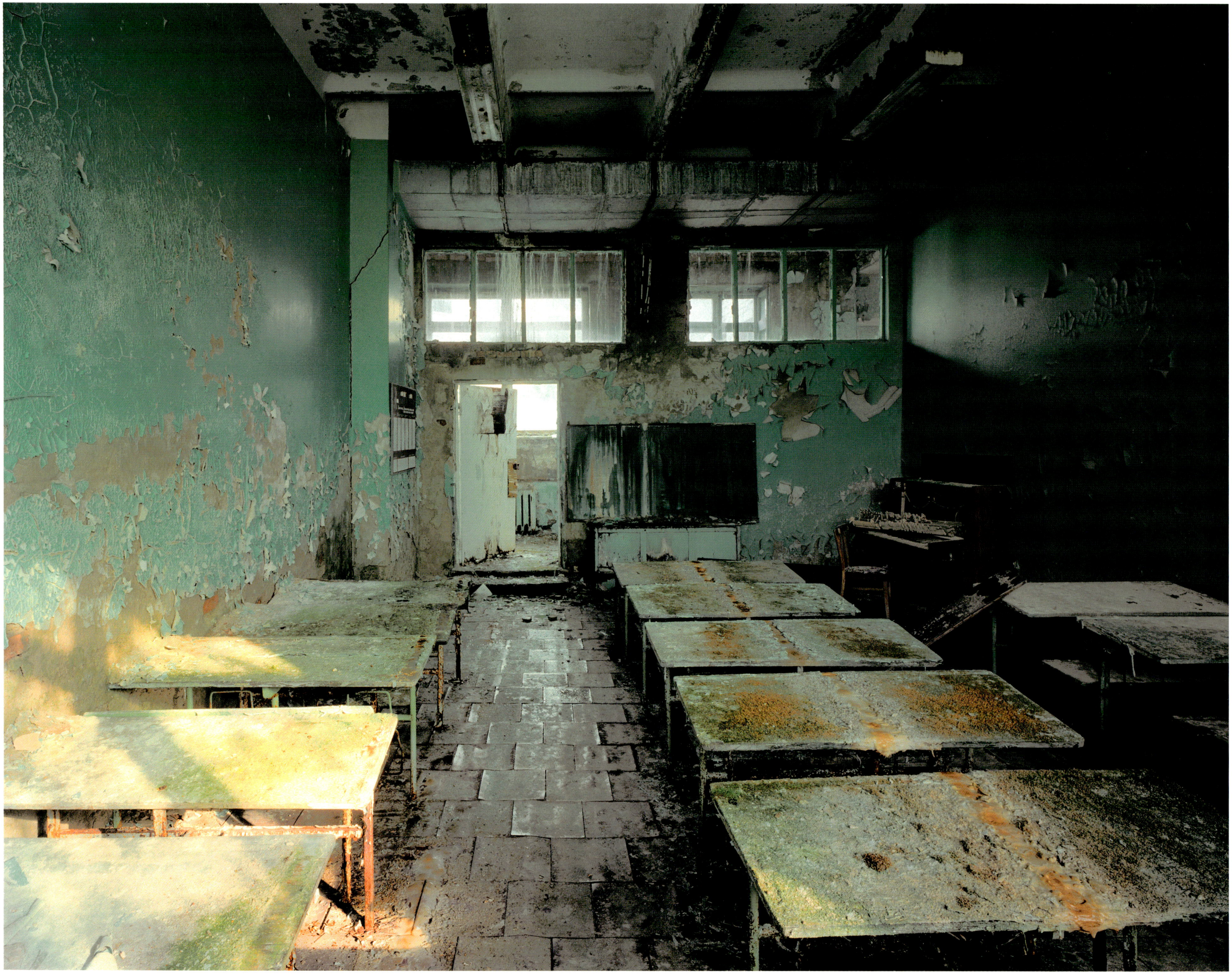

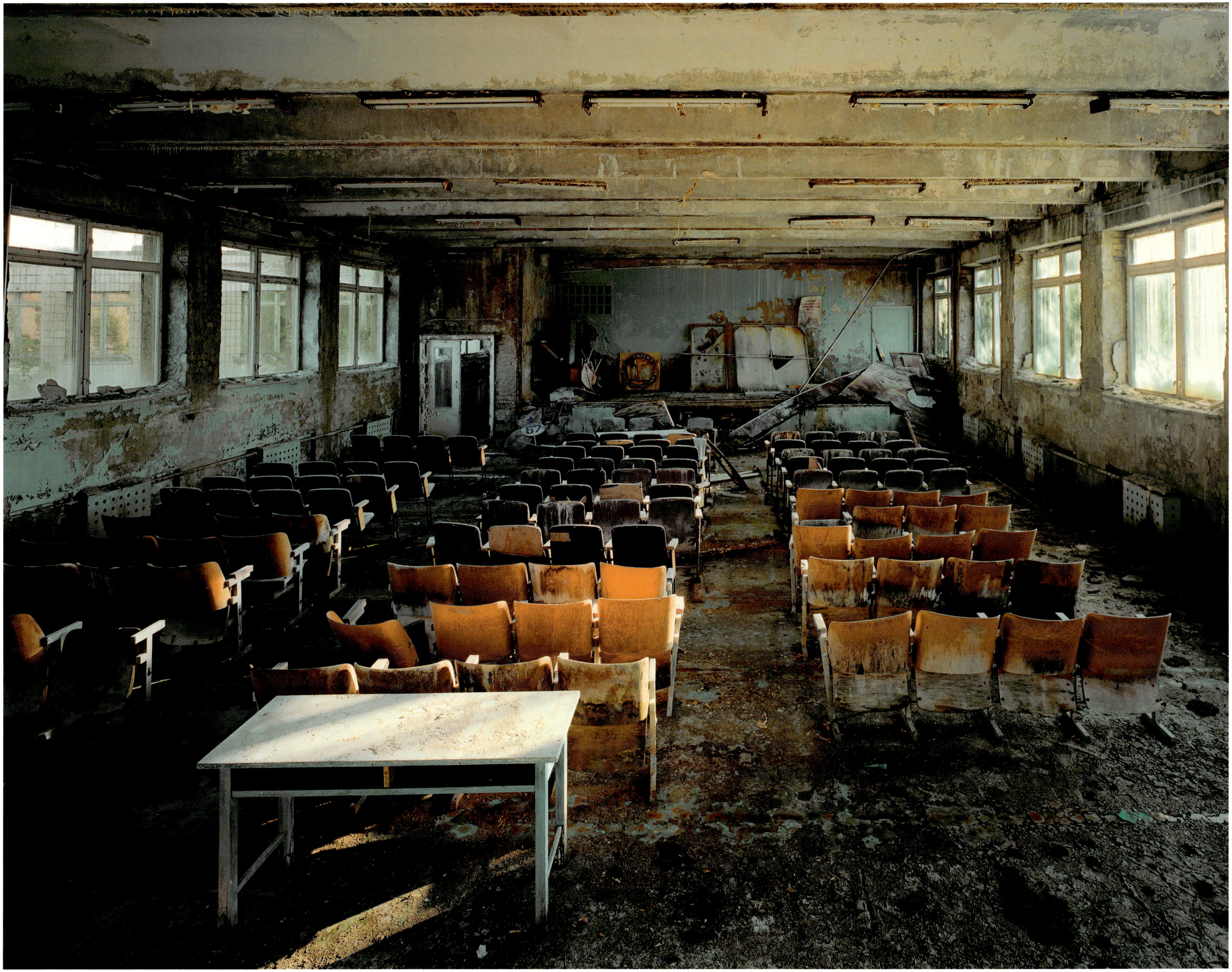

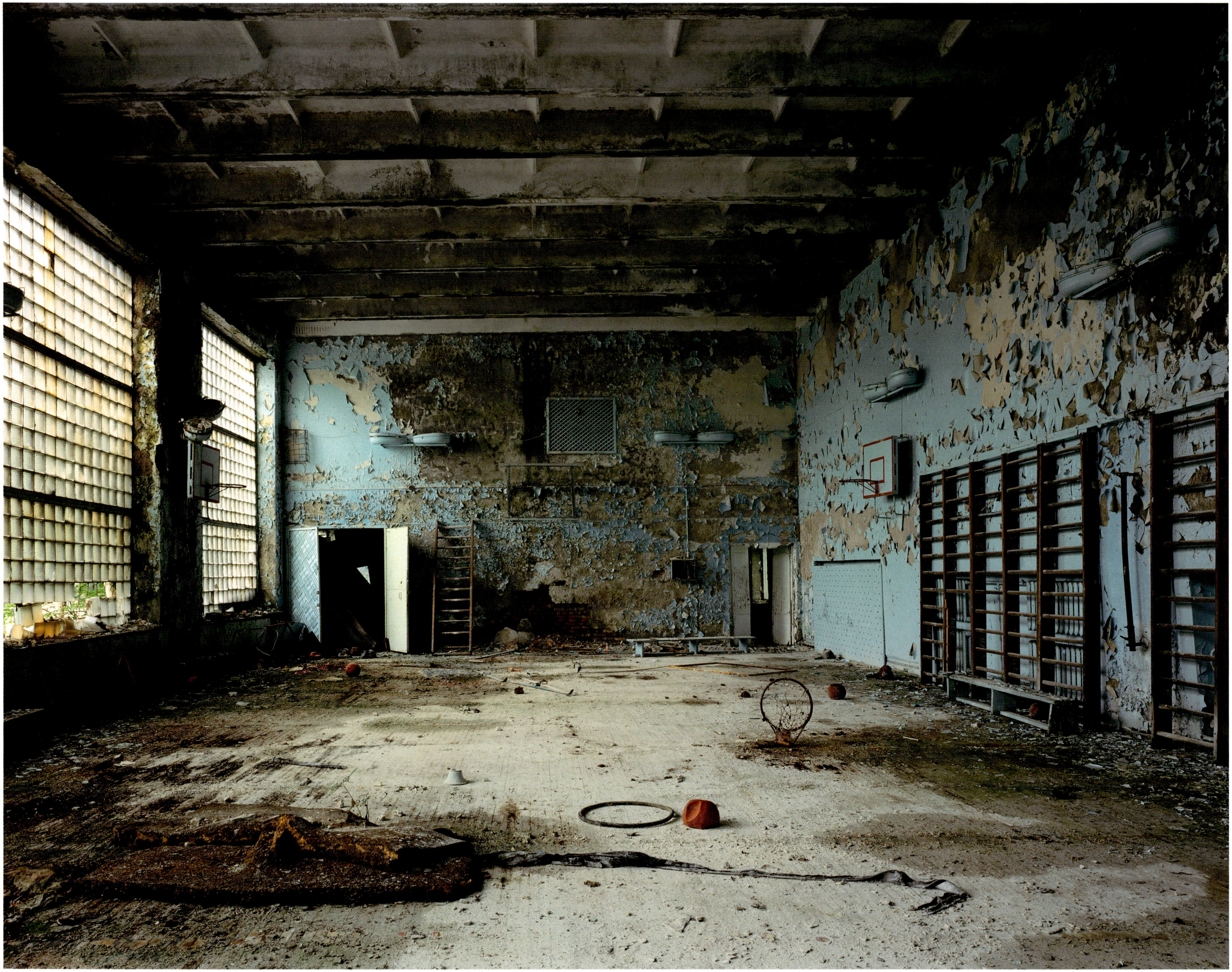

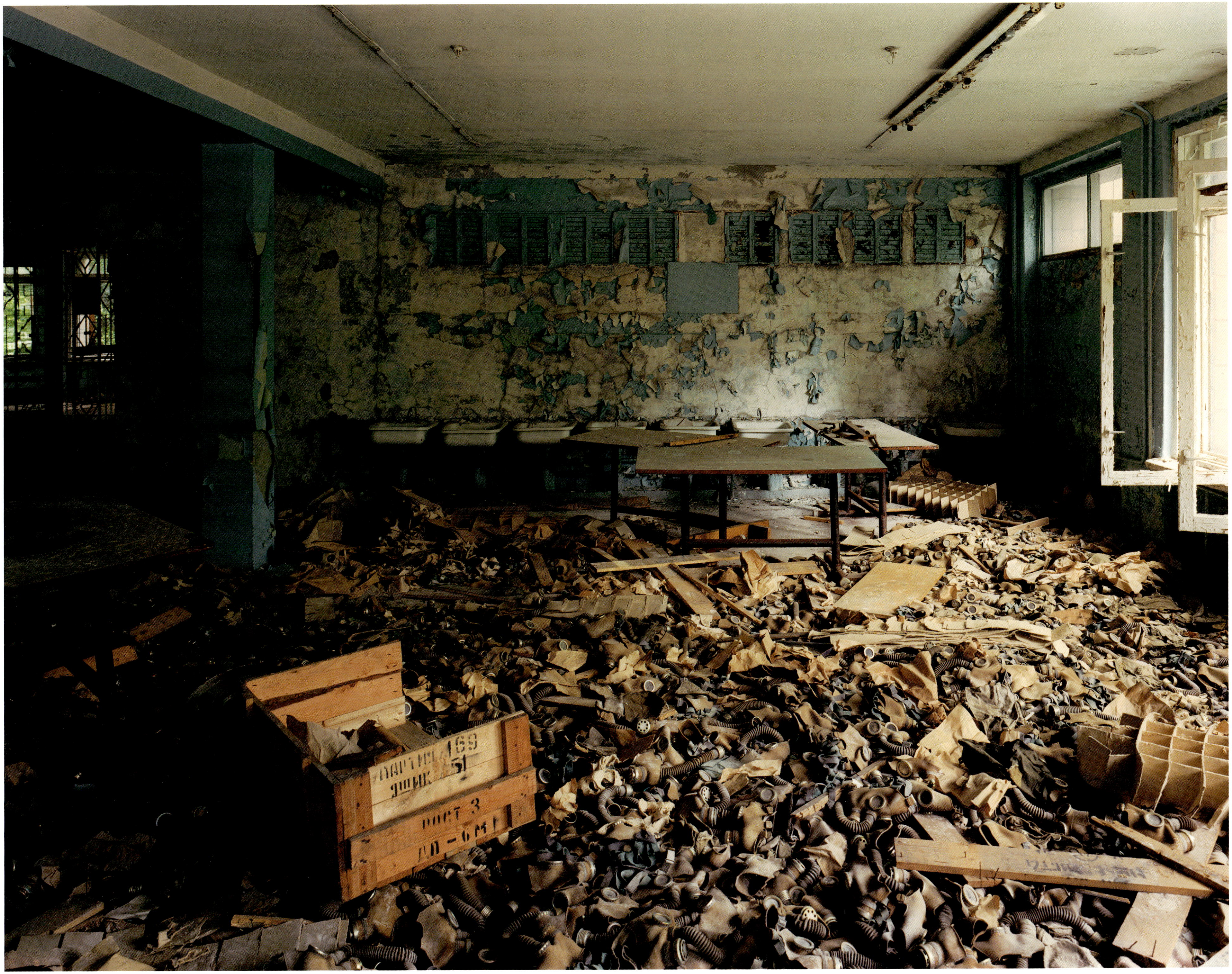
ПАРТИЯ 169
ЯЩИК 51
ДП-6М

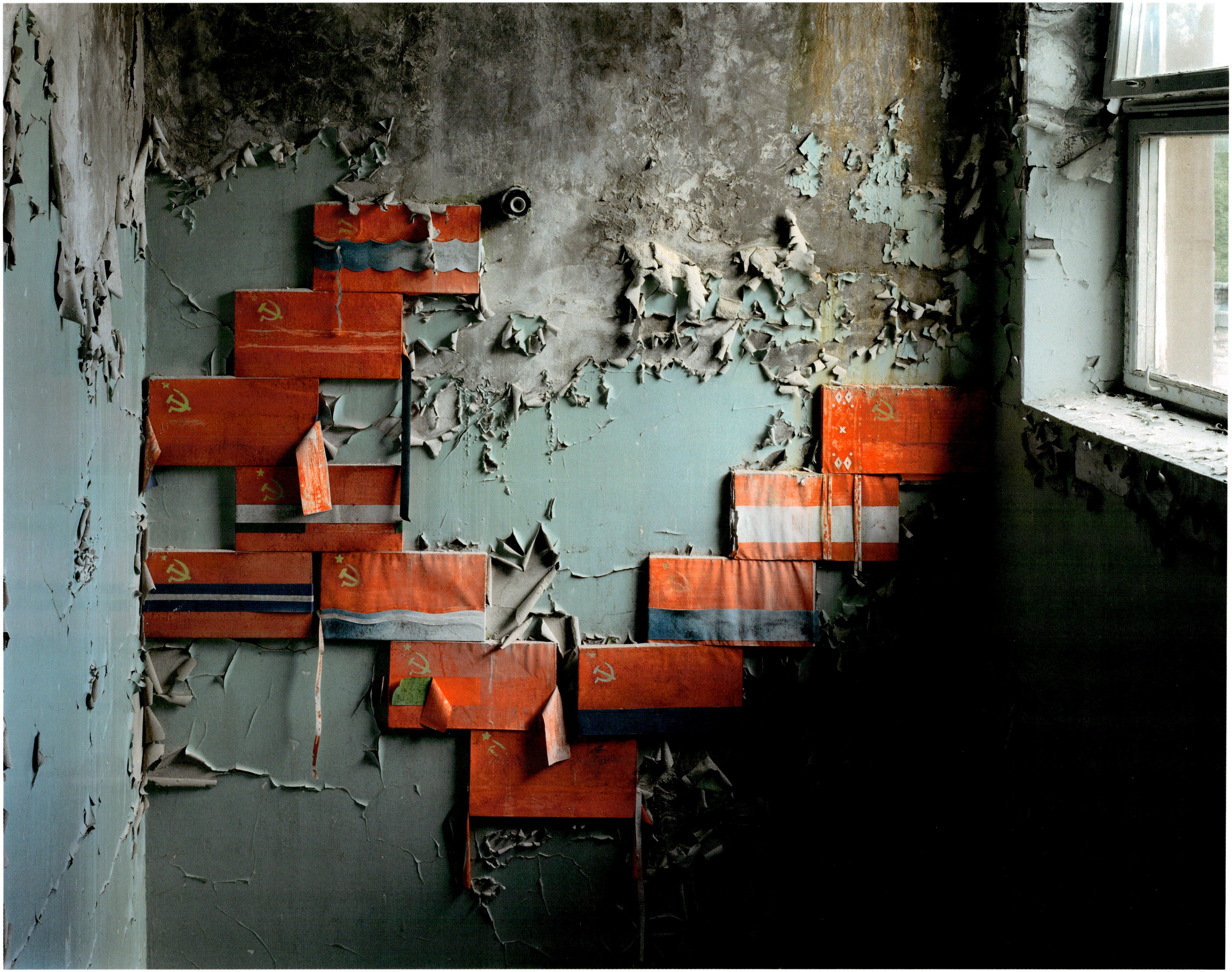

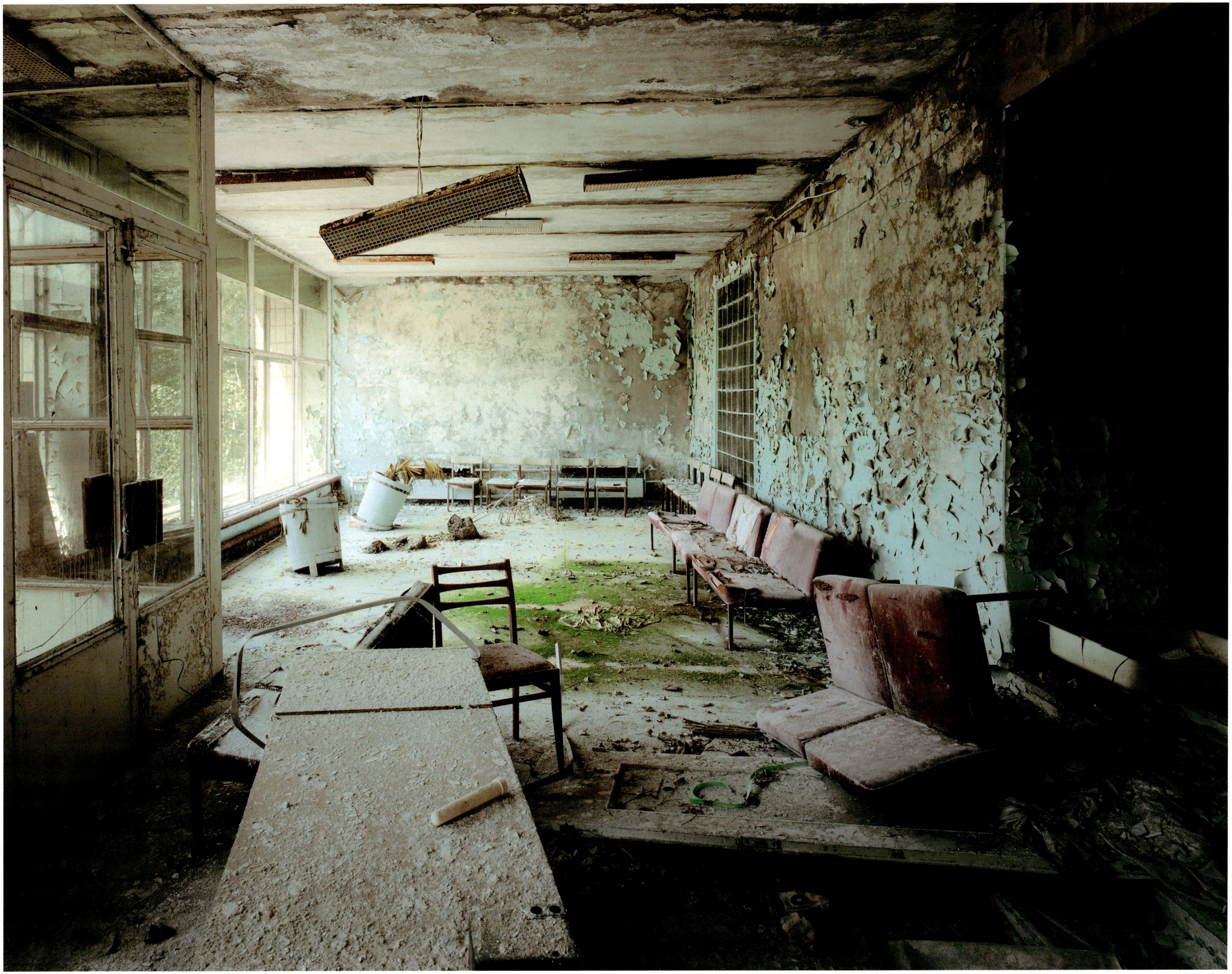

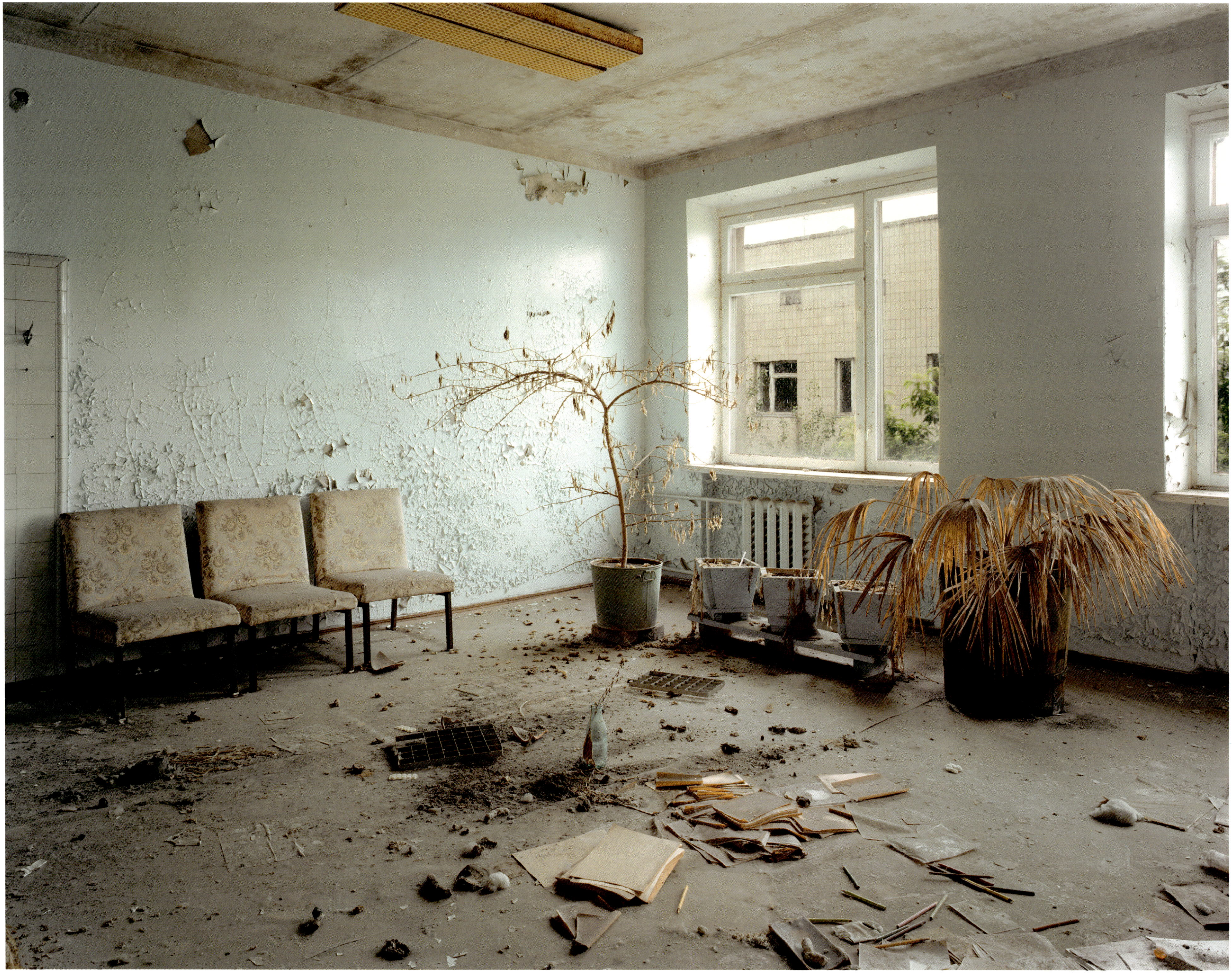

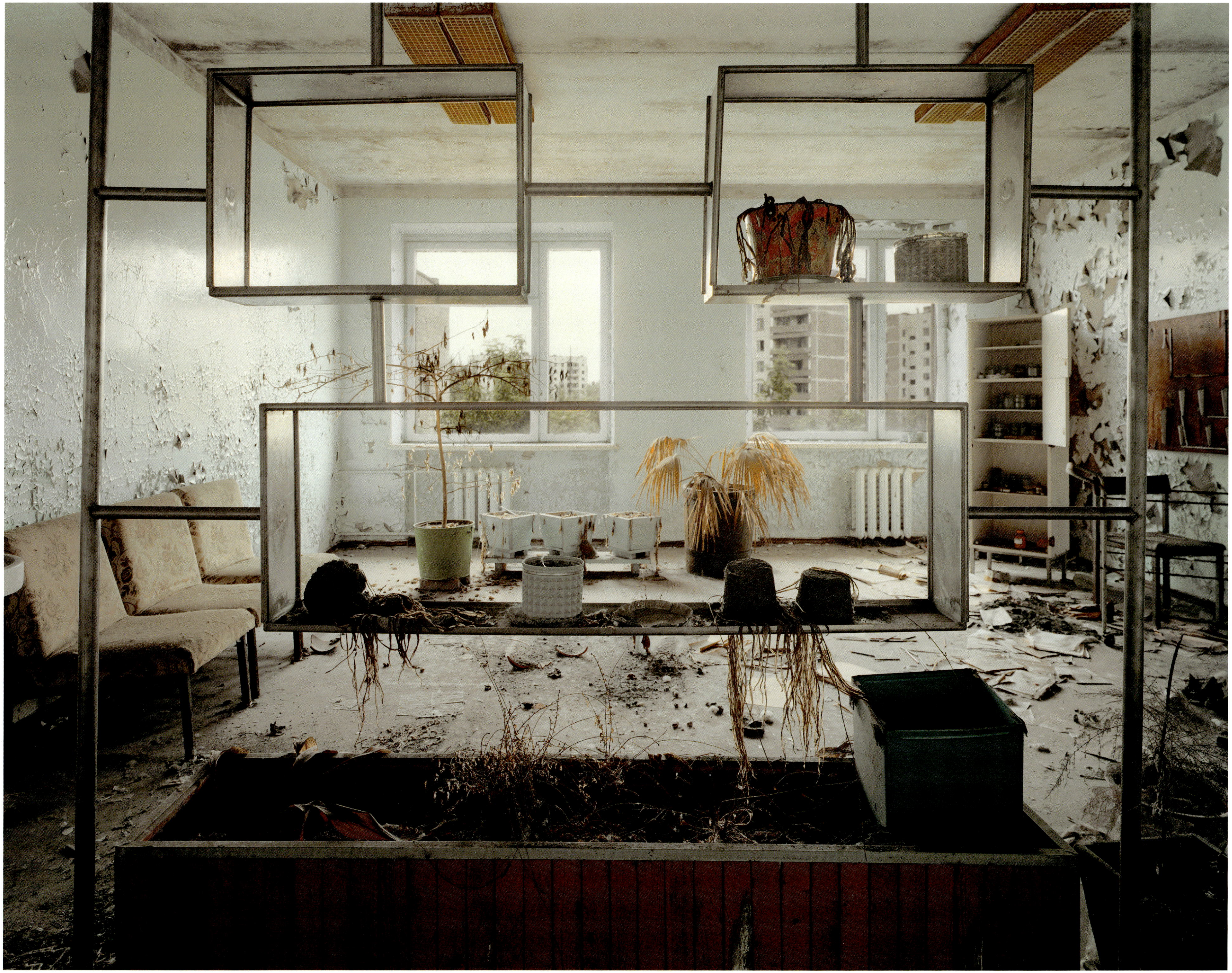

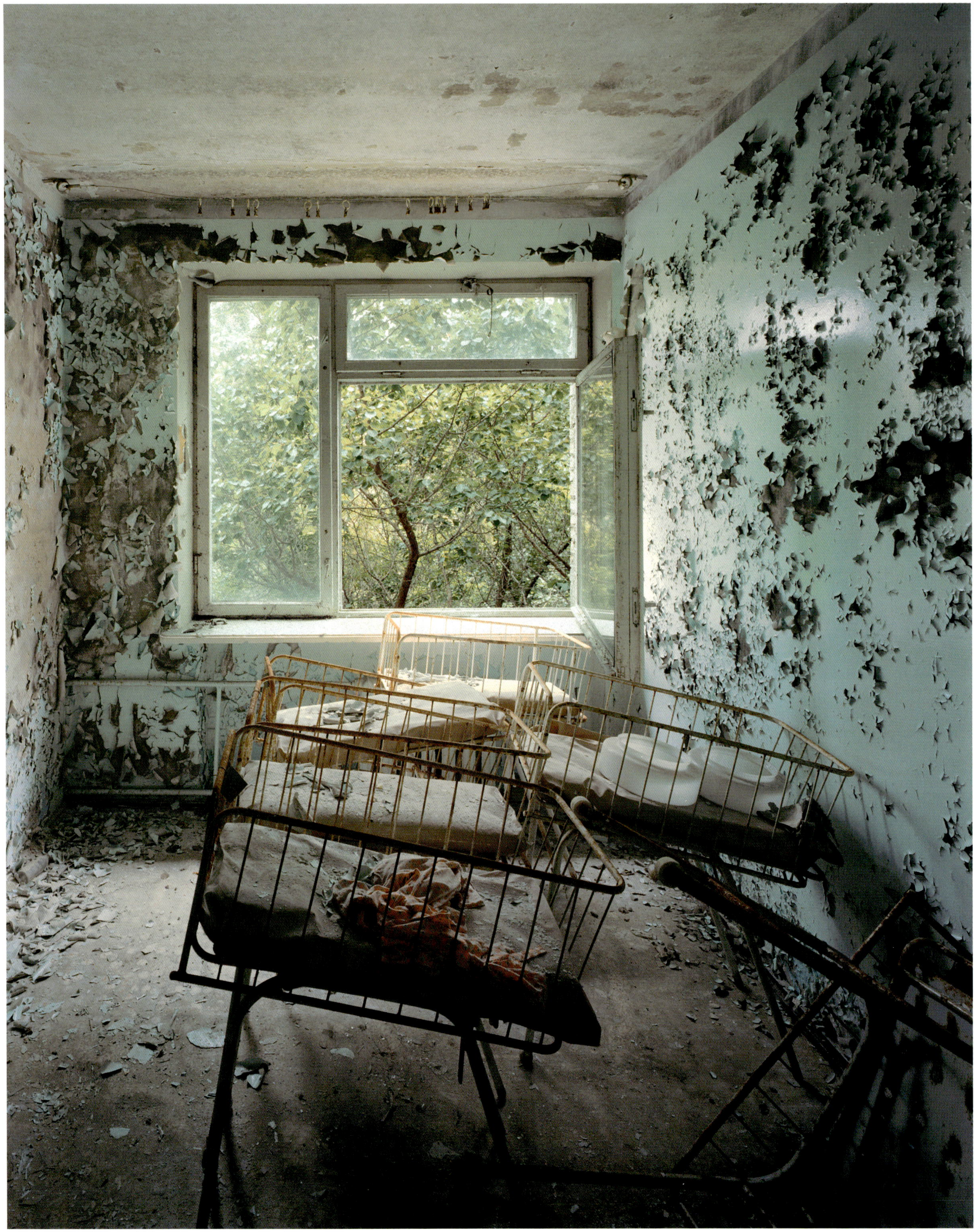

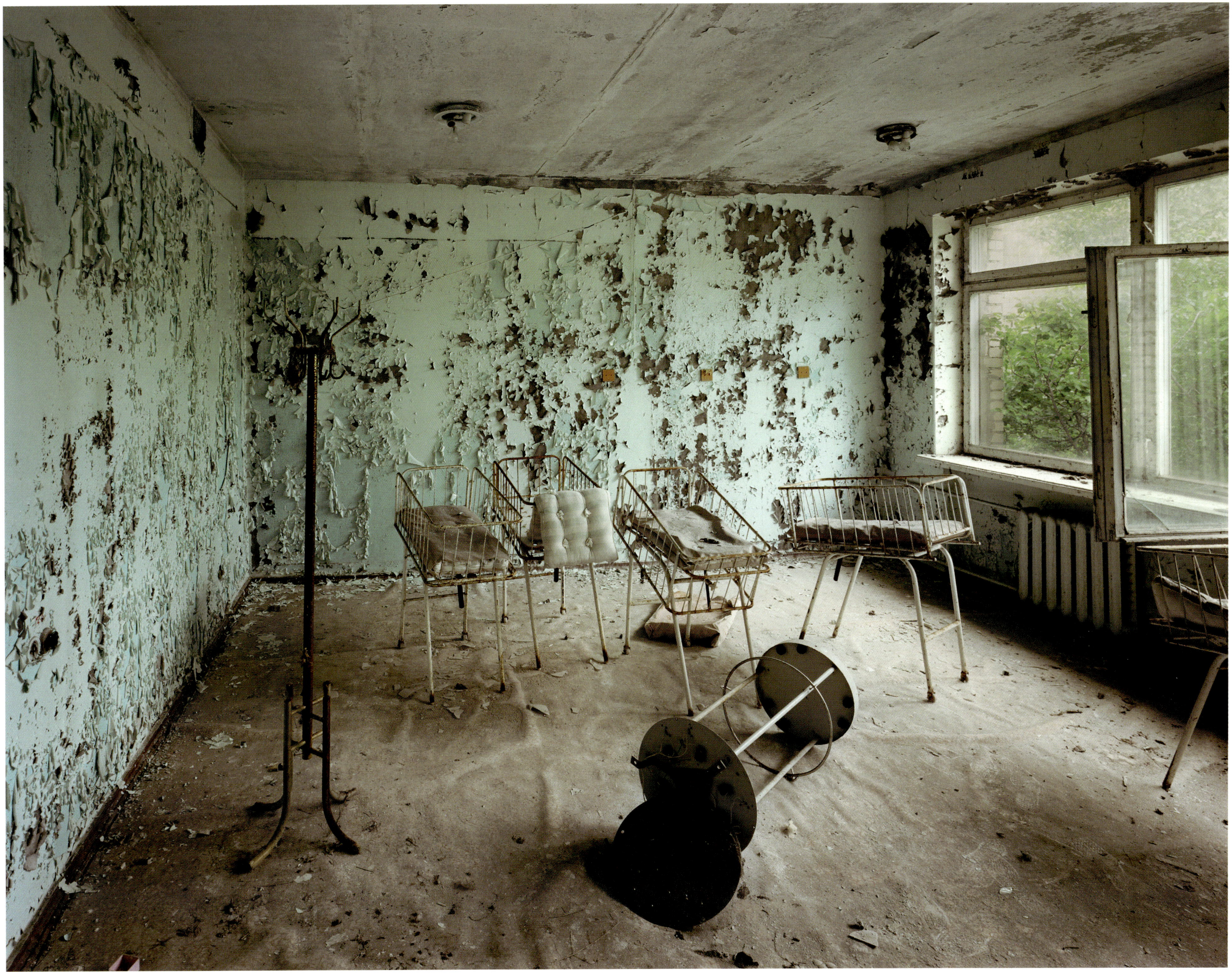

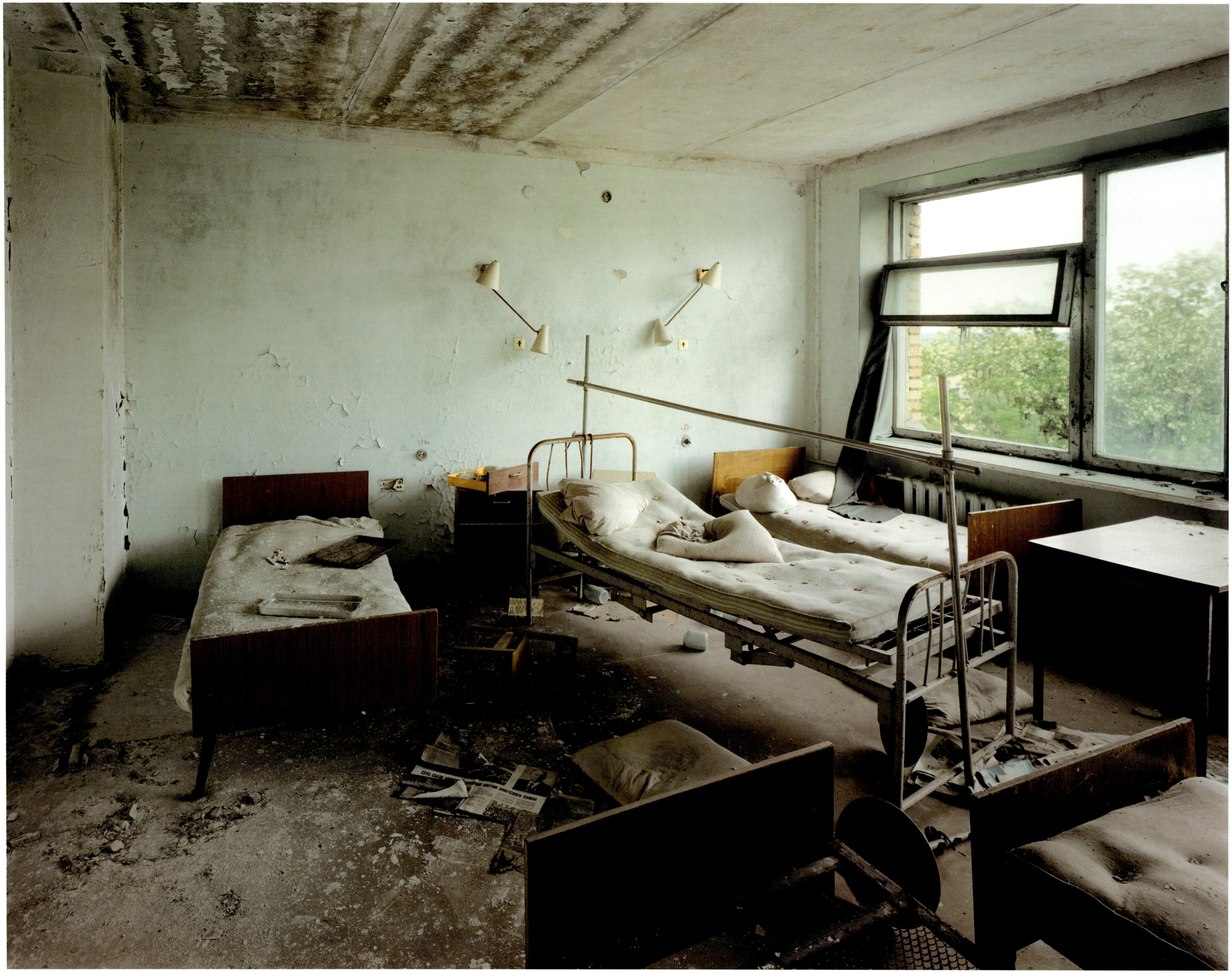
XXVII СЪЕЗД
ПАРТИИ СОВЕТСКОГО СОЮЗА

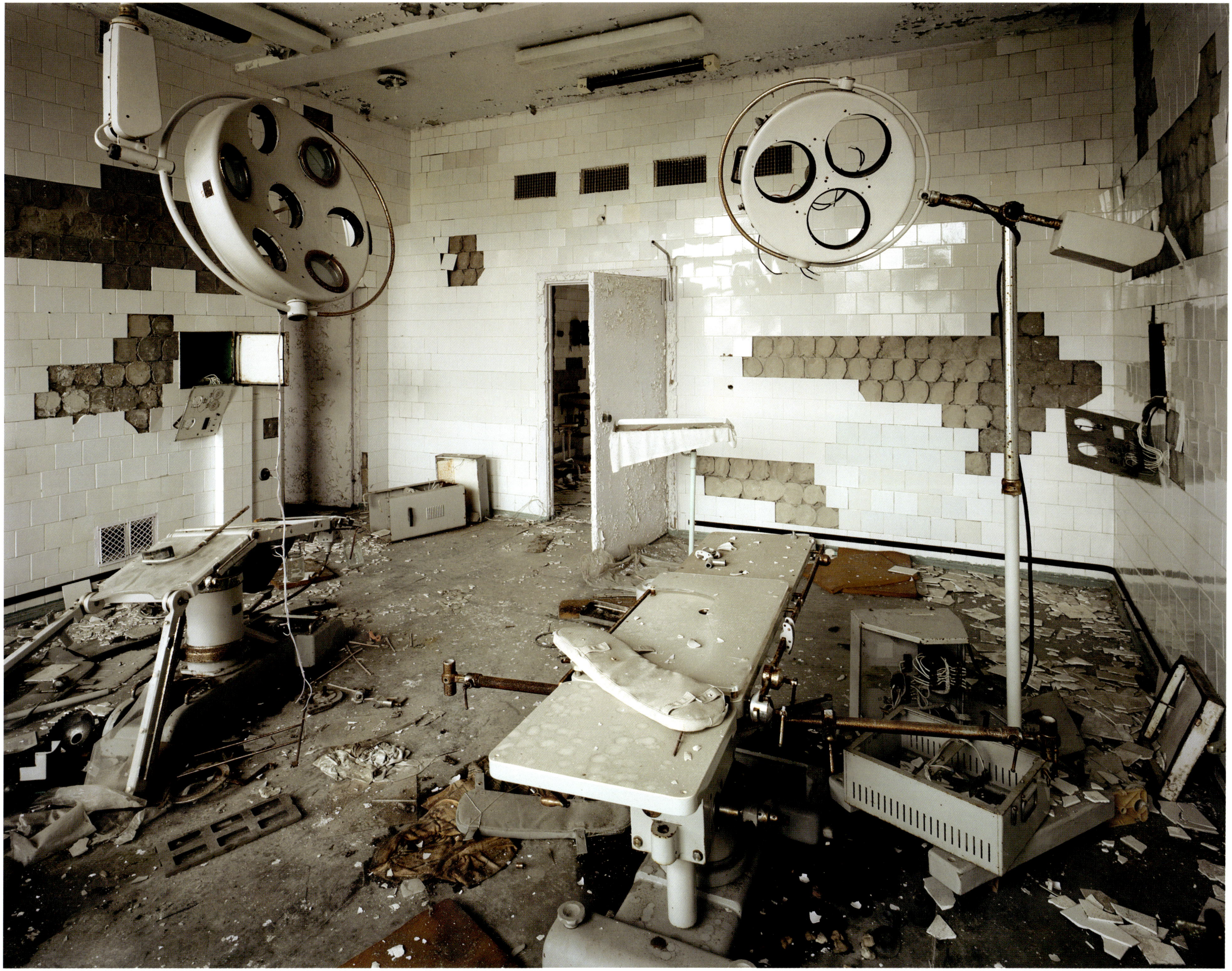

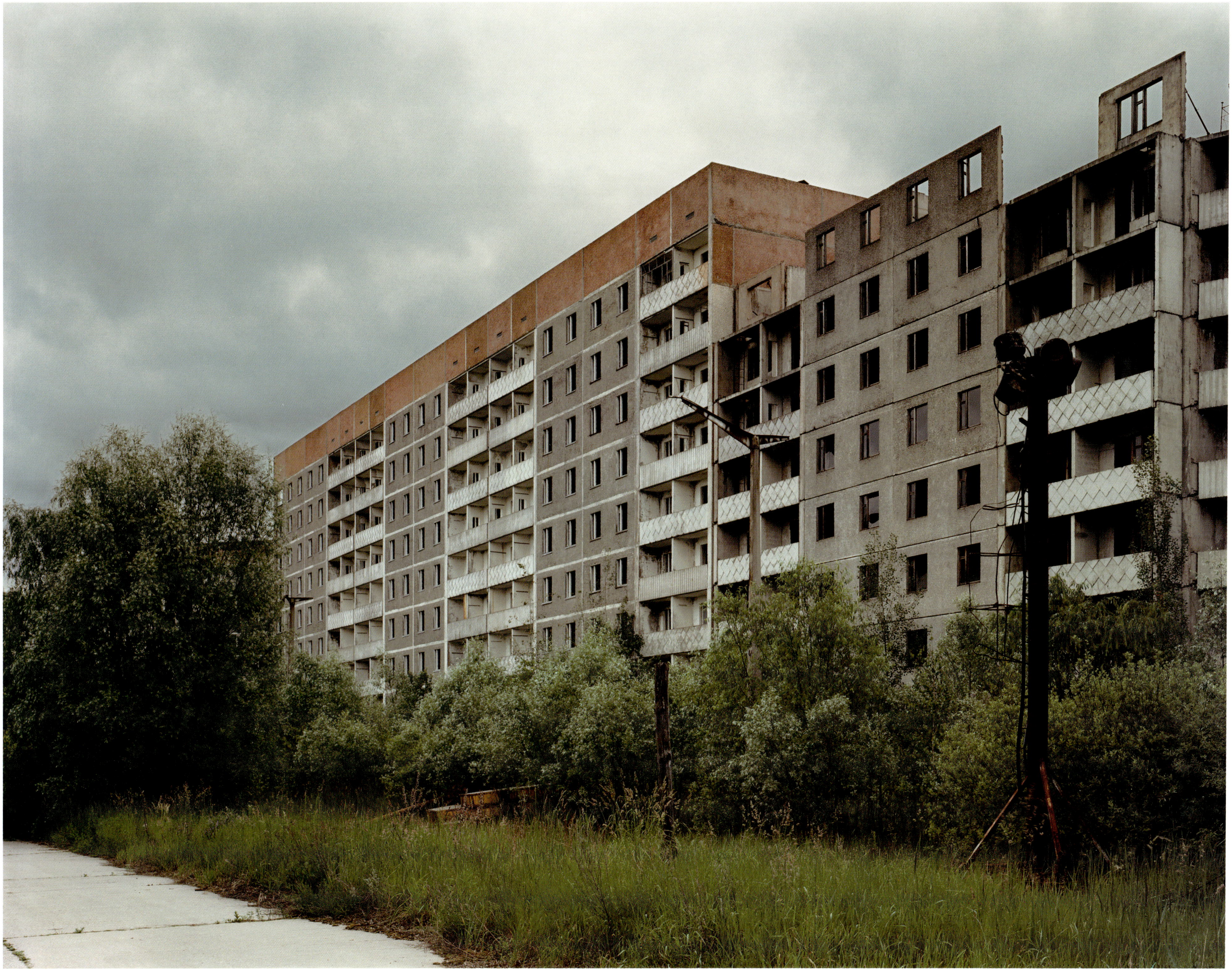

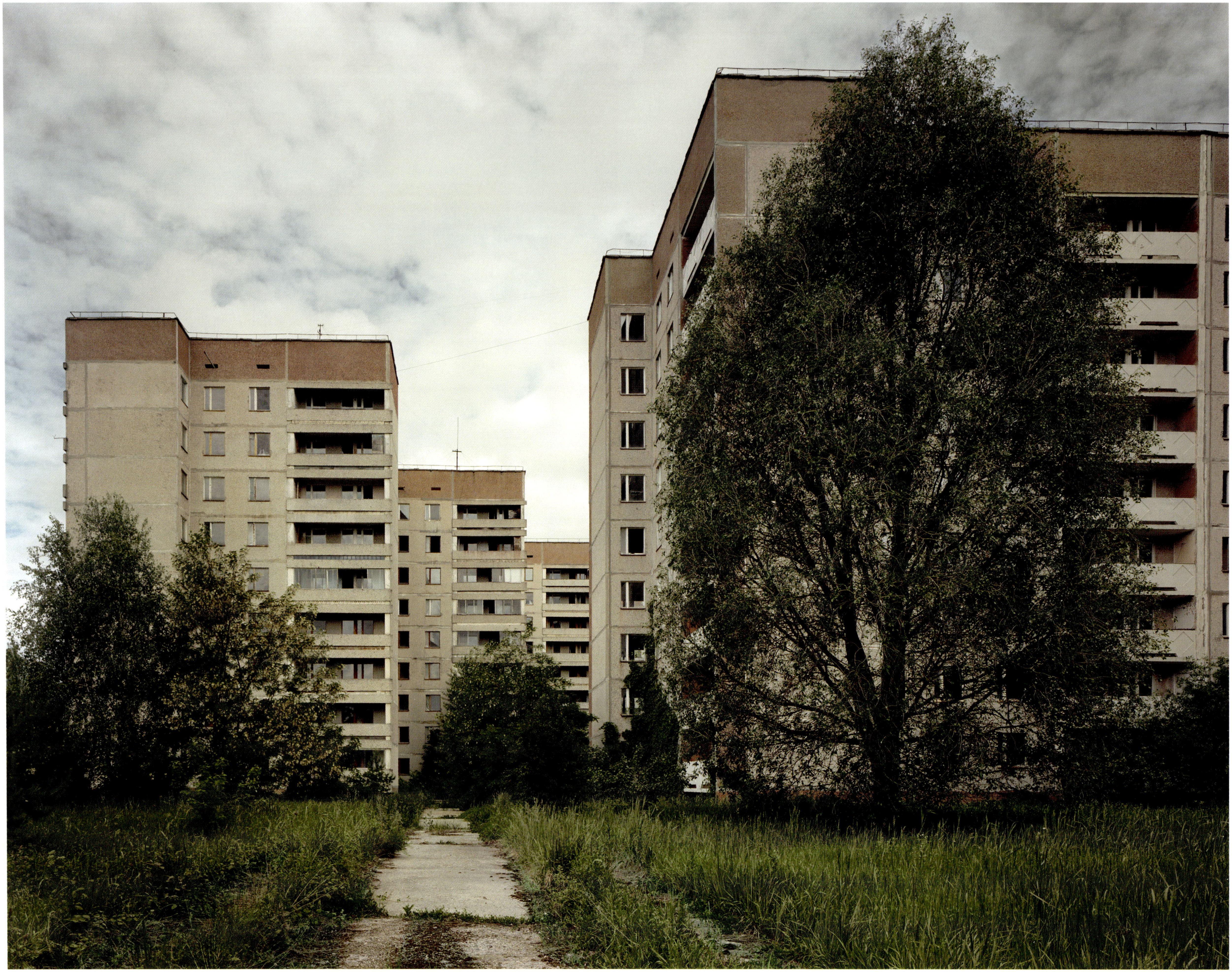

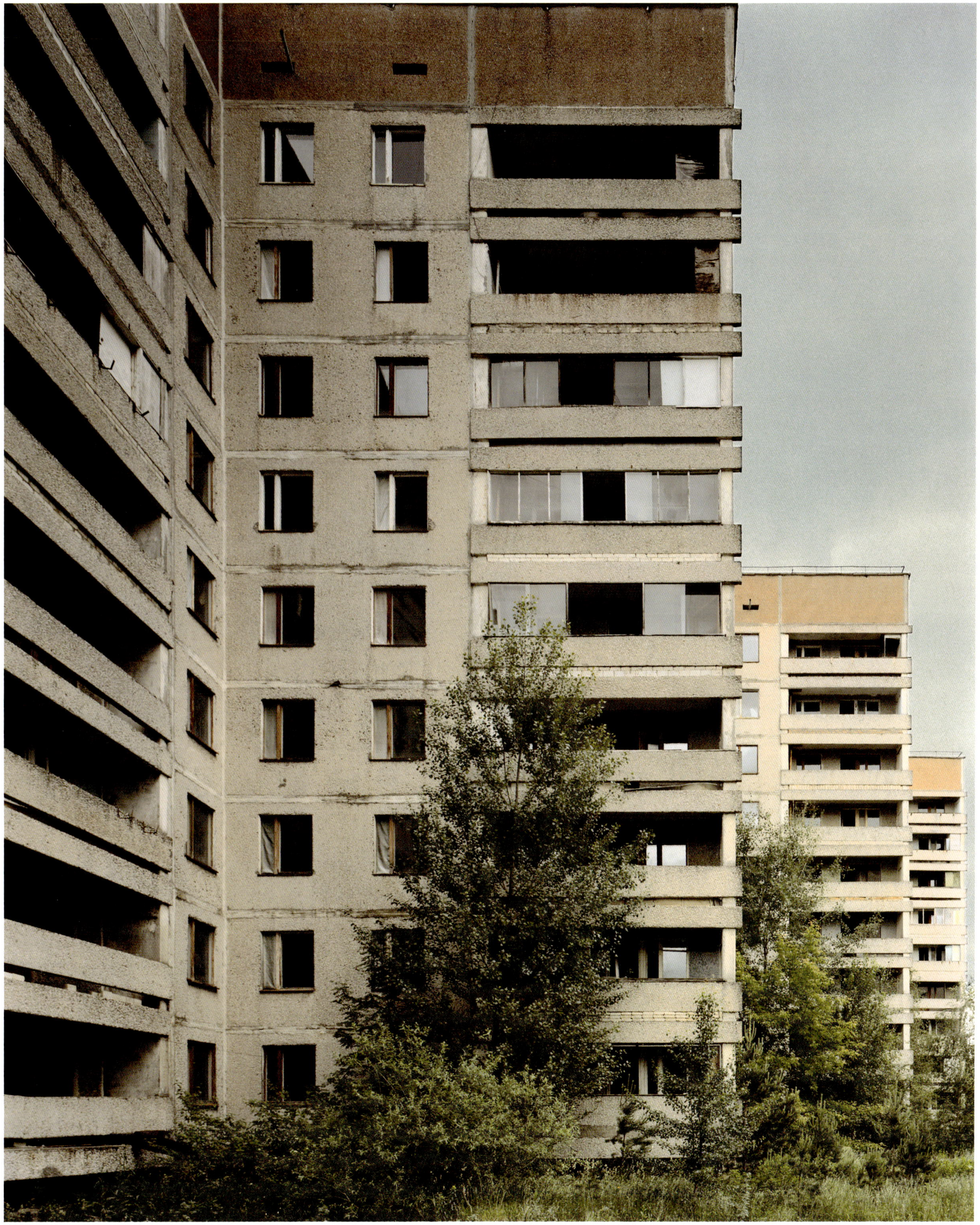

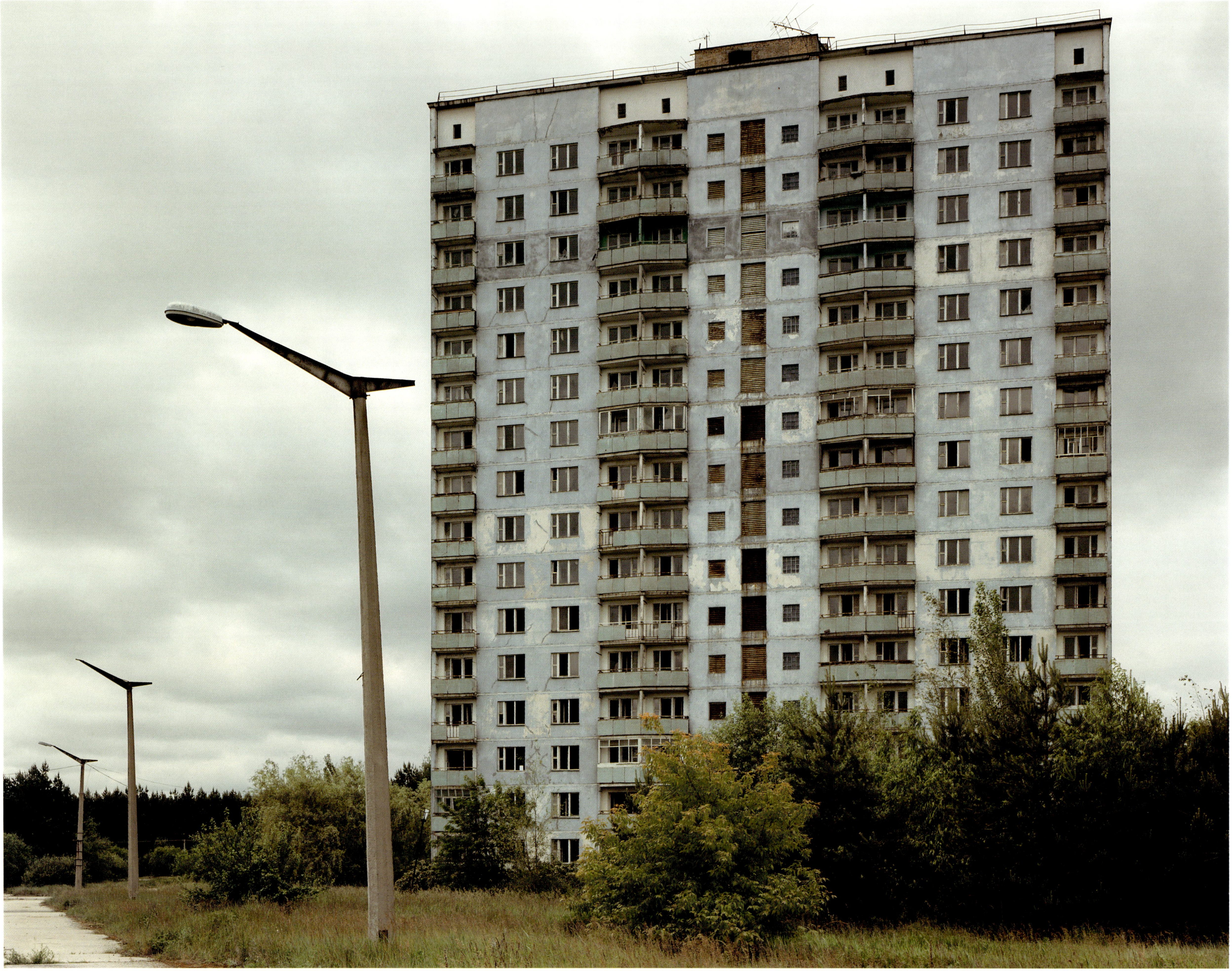

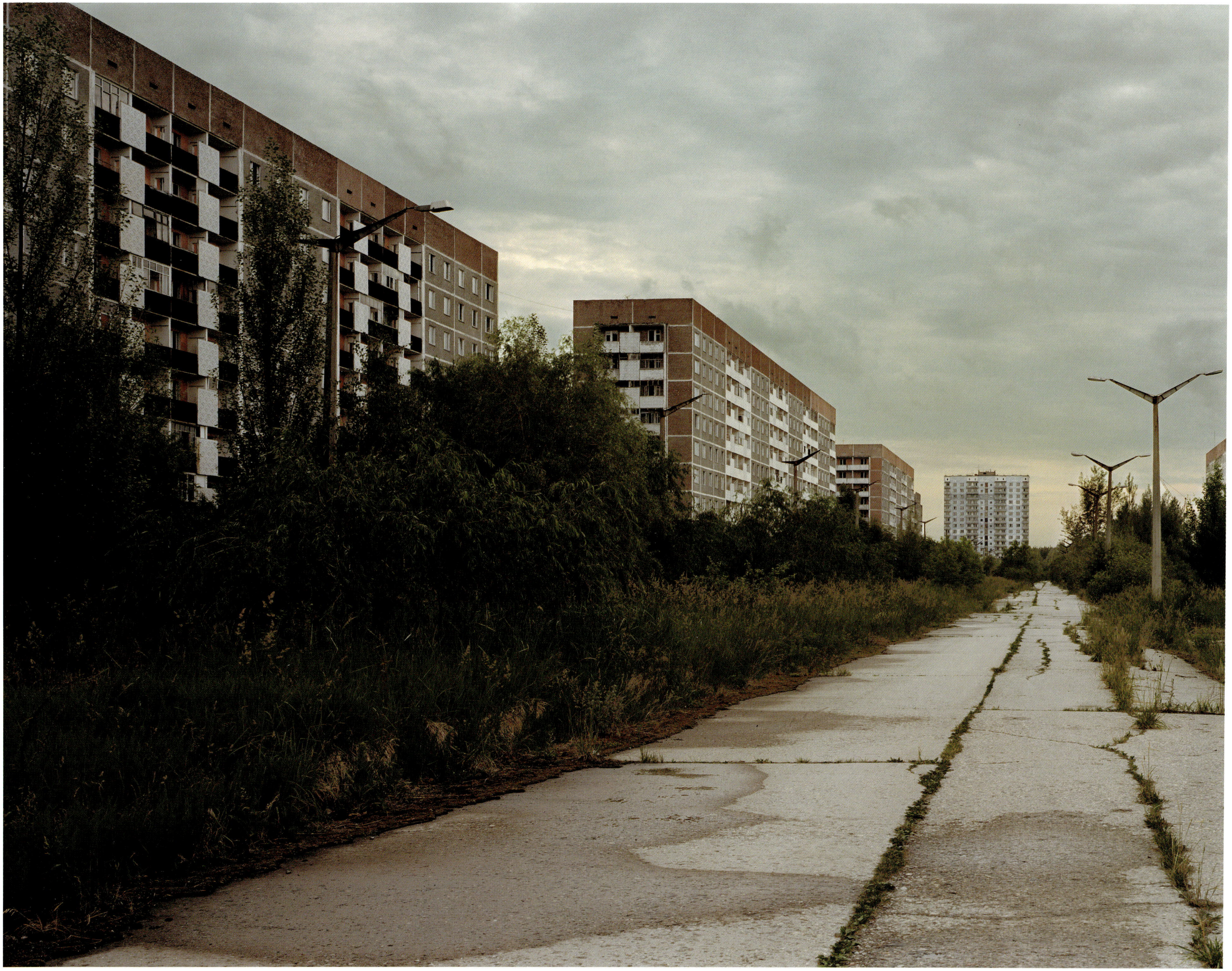

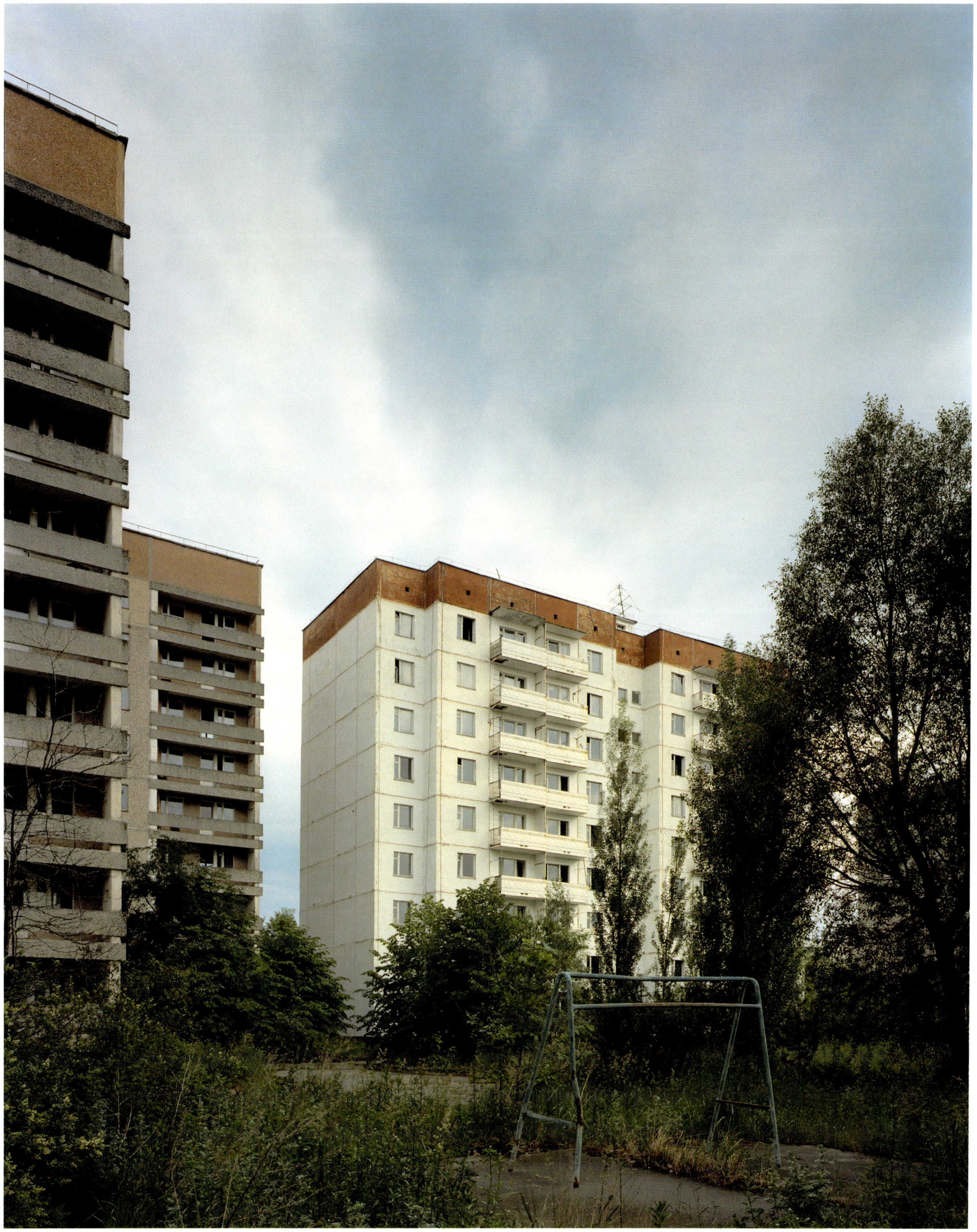

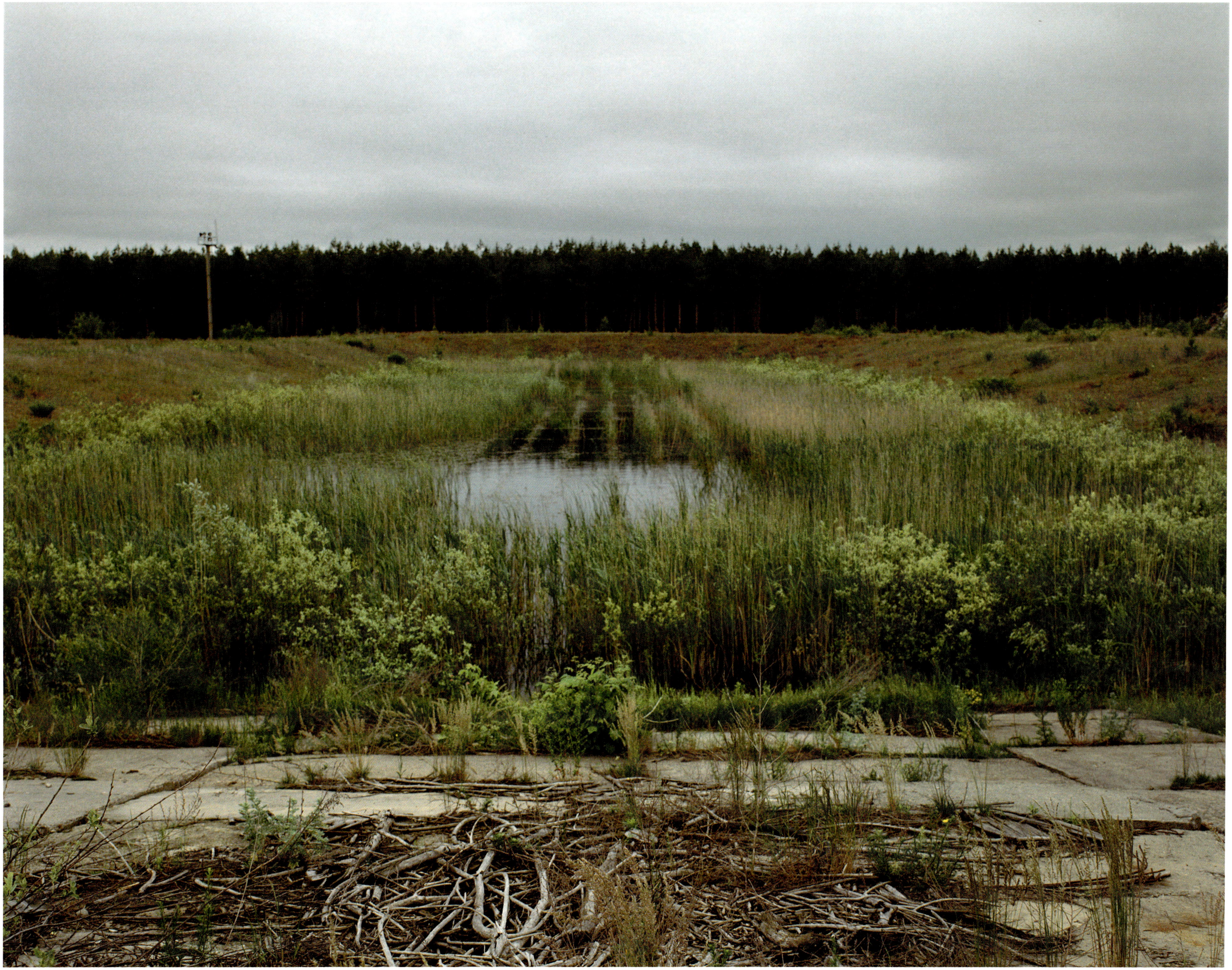

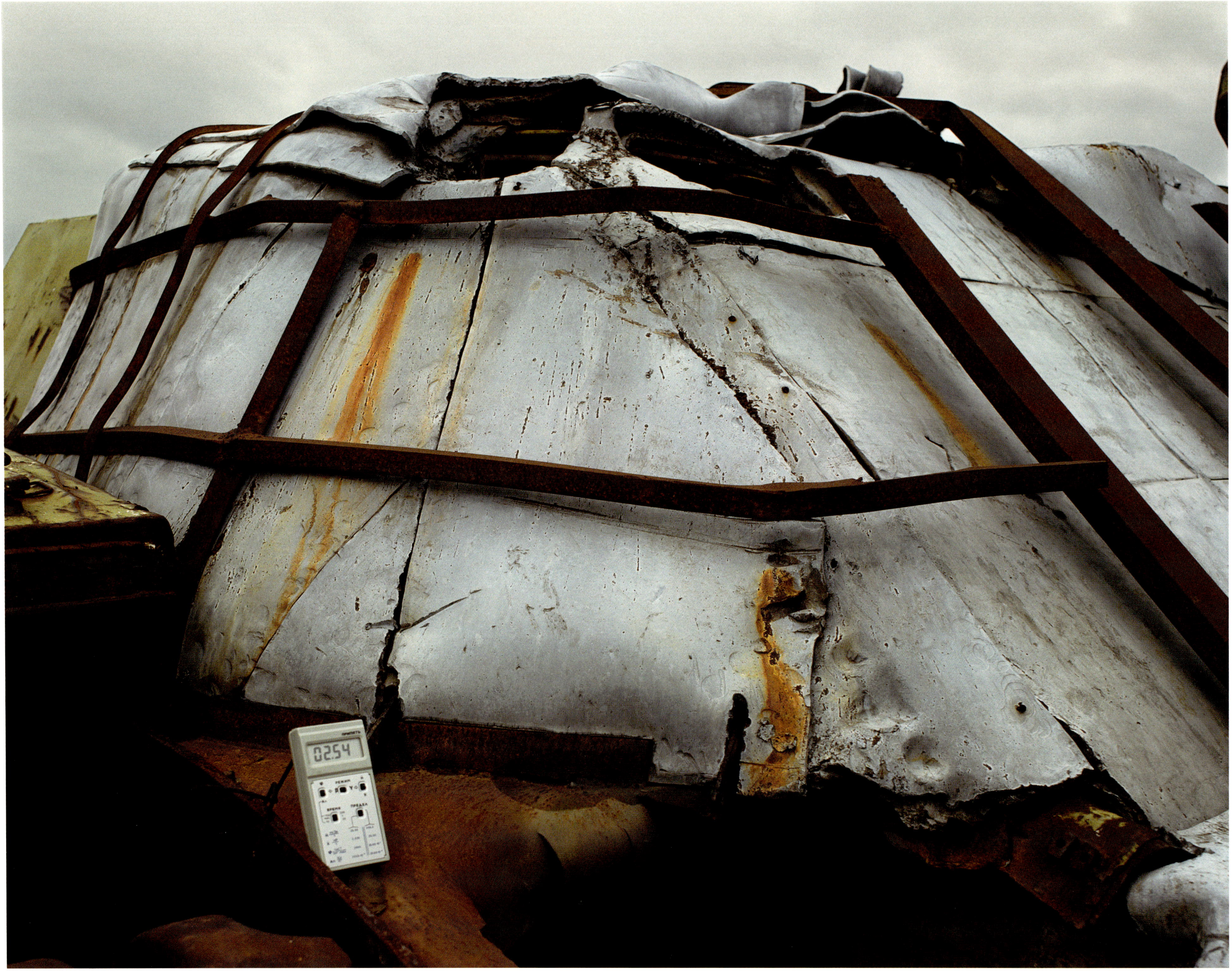
02.54
РЕЖИМ
ВРЕМЯ
ПРЕДЕЛ

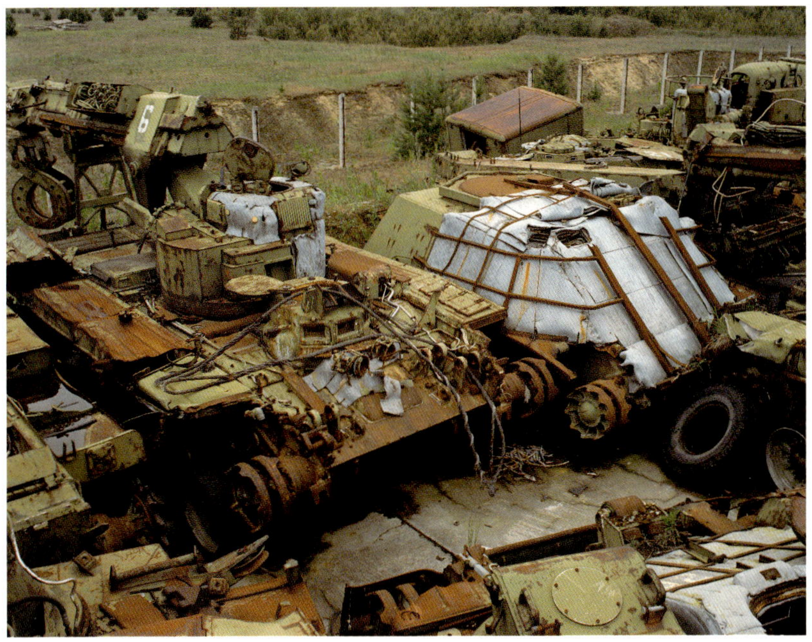
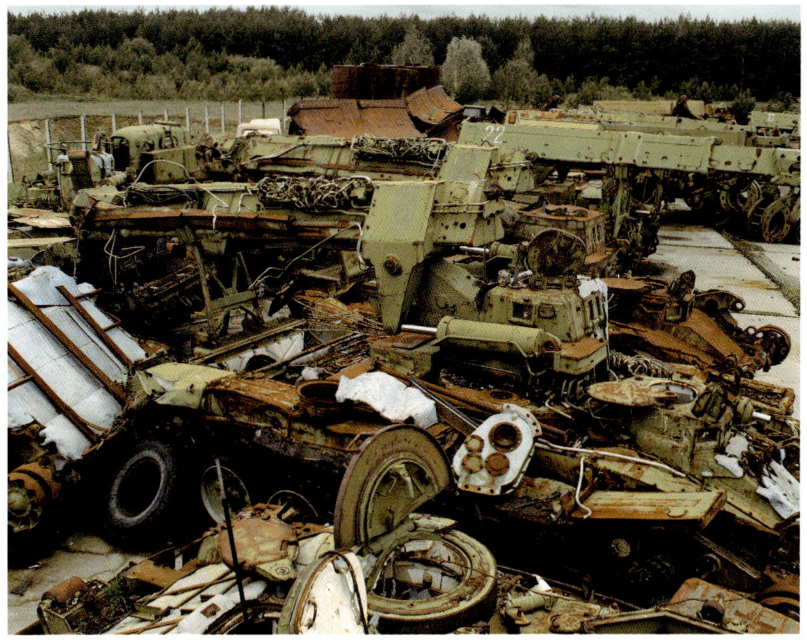
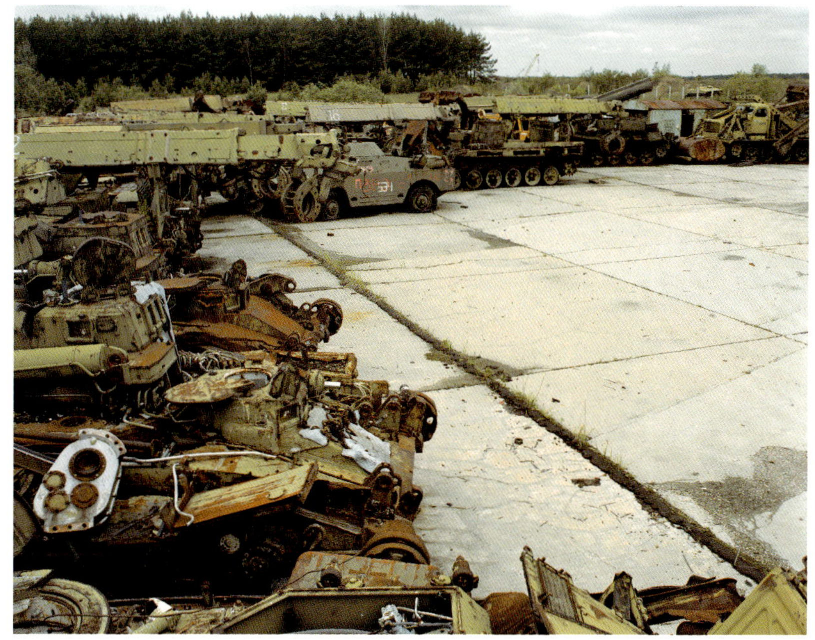

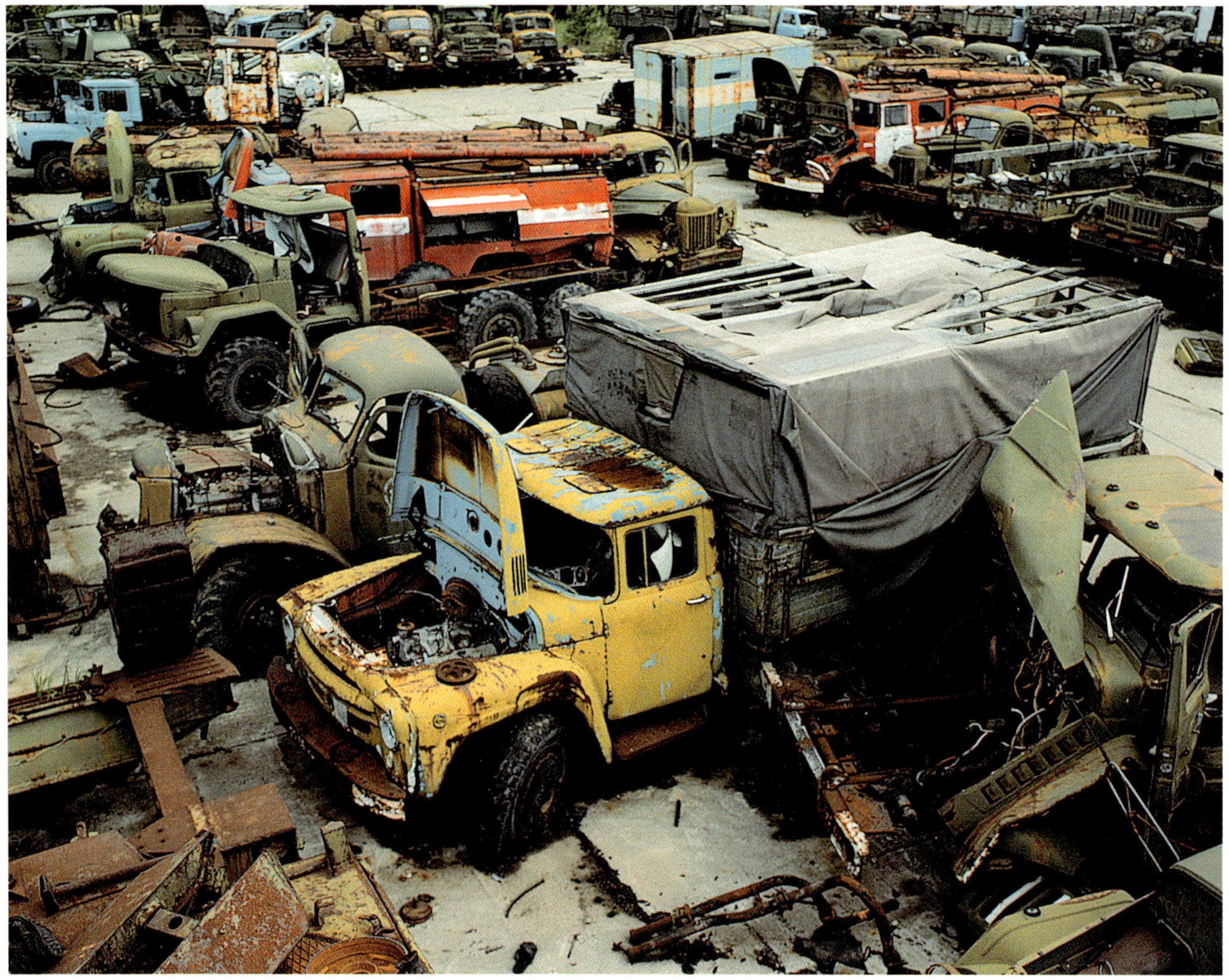

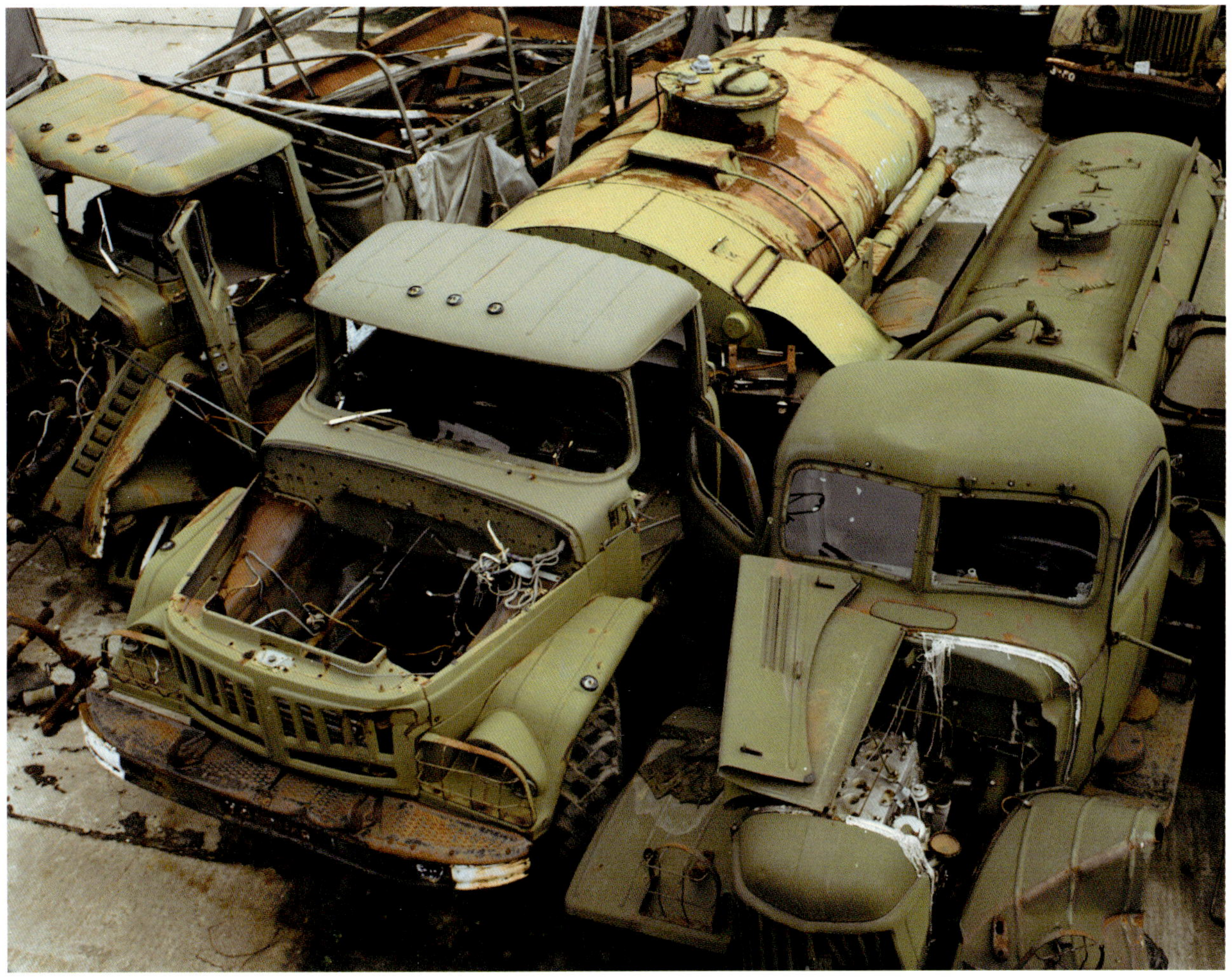

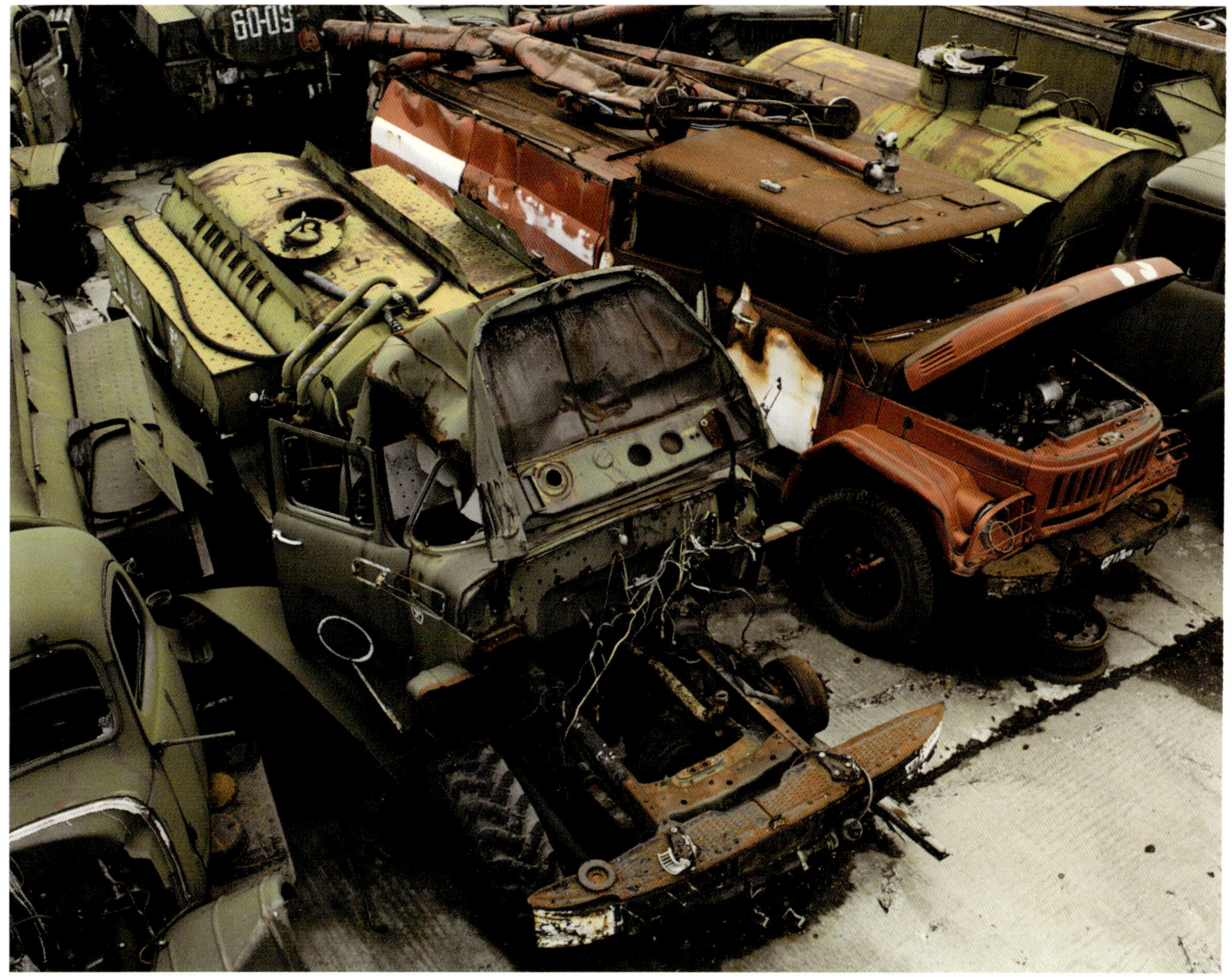
60-09

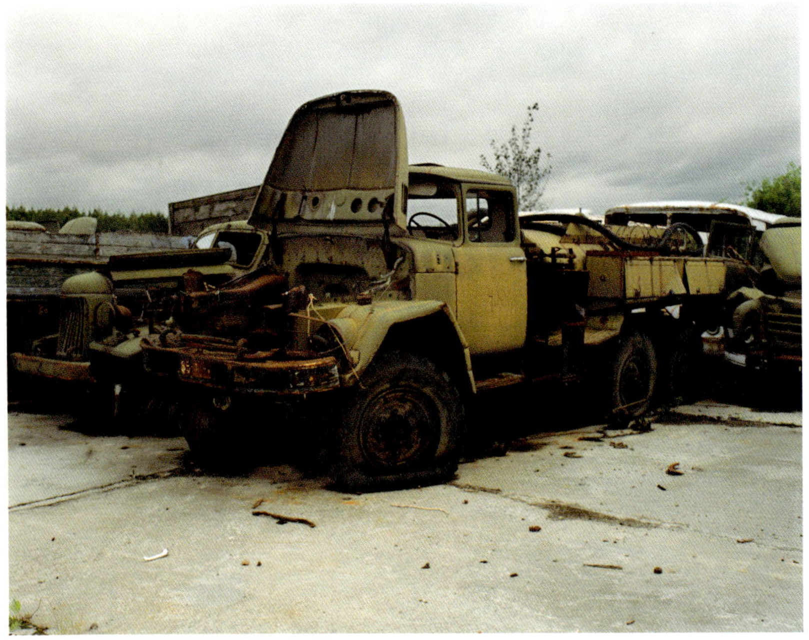
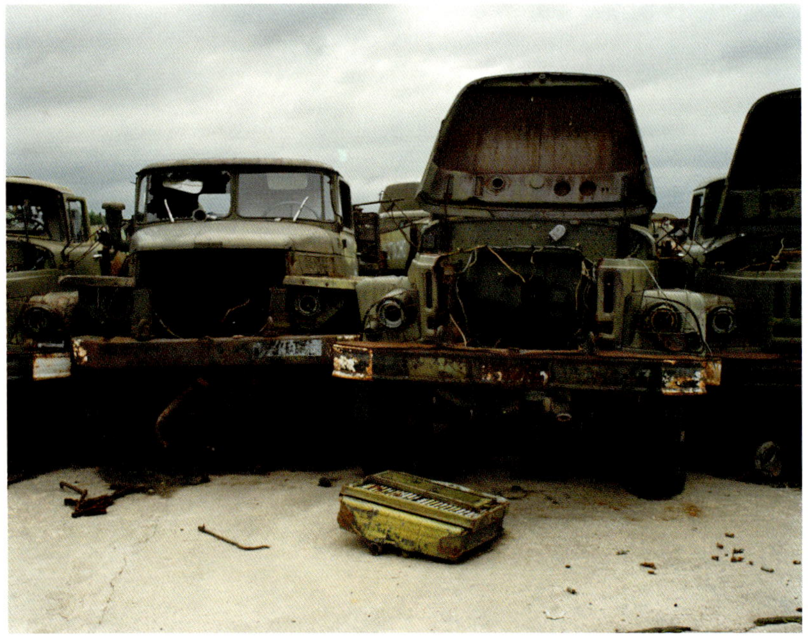
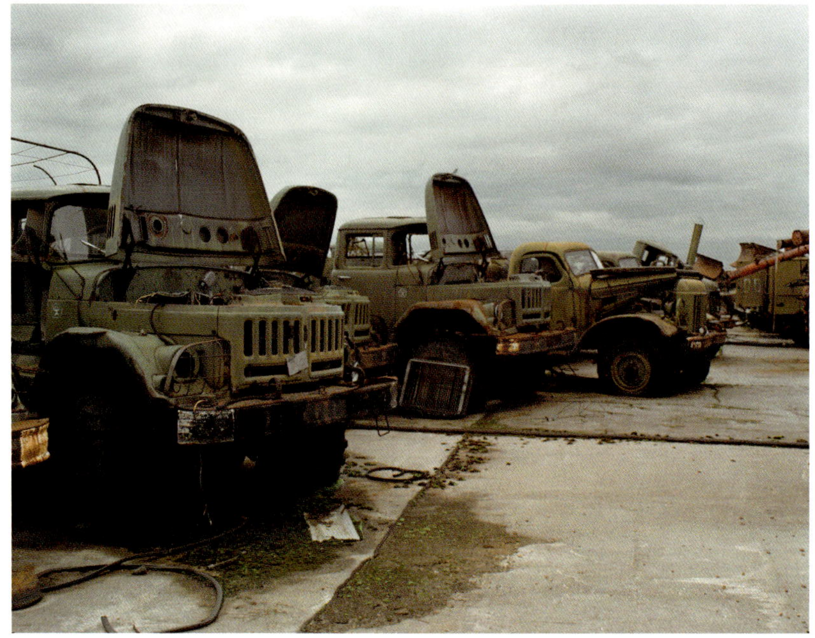
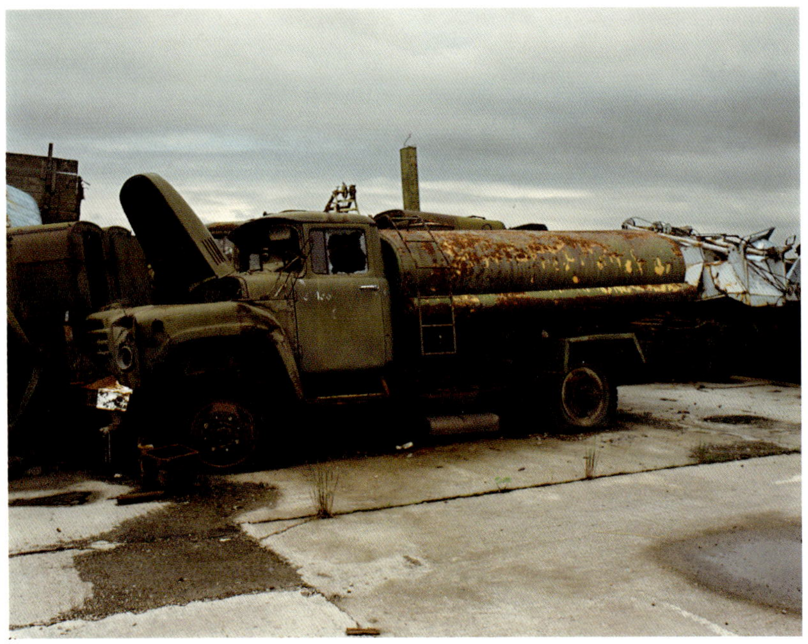
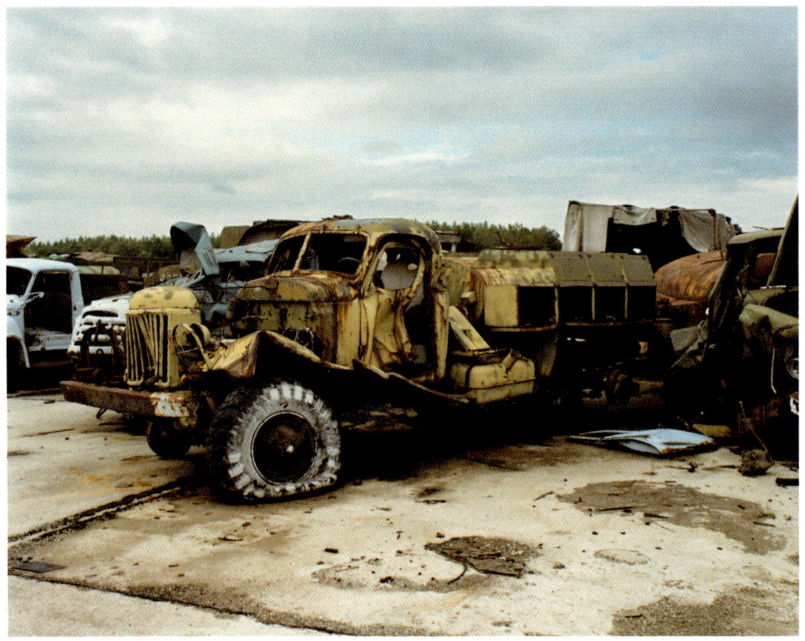
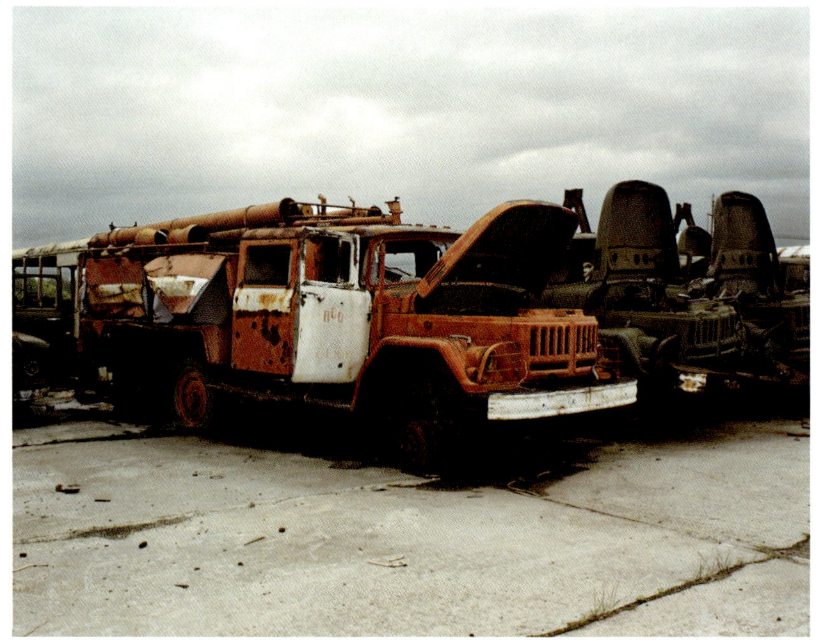

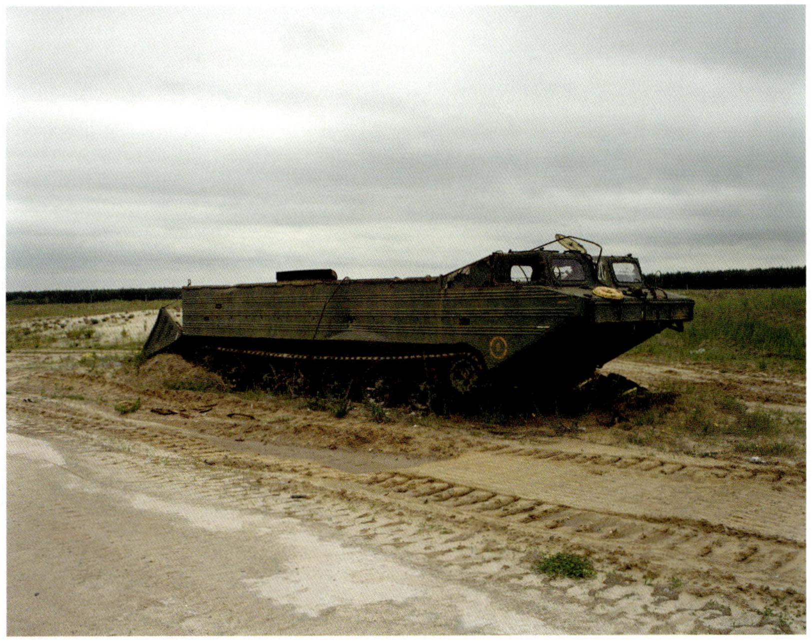
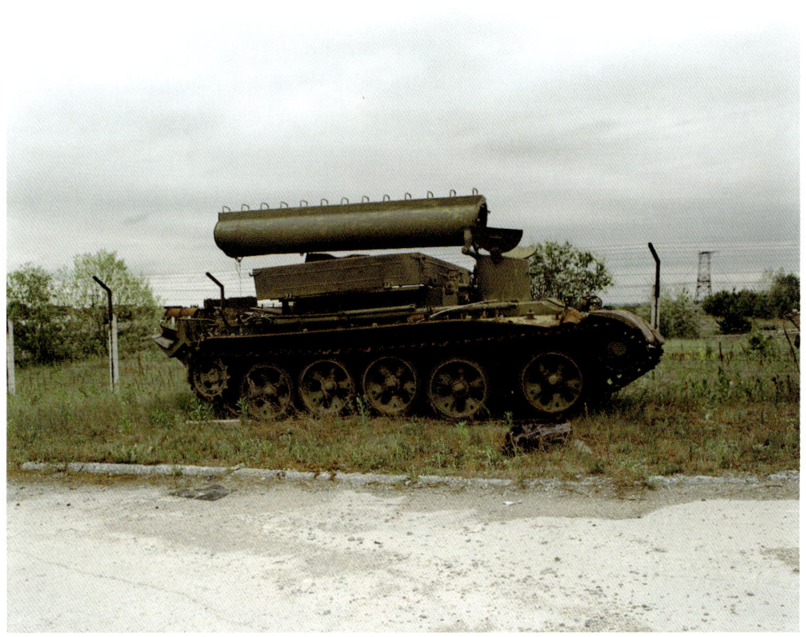
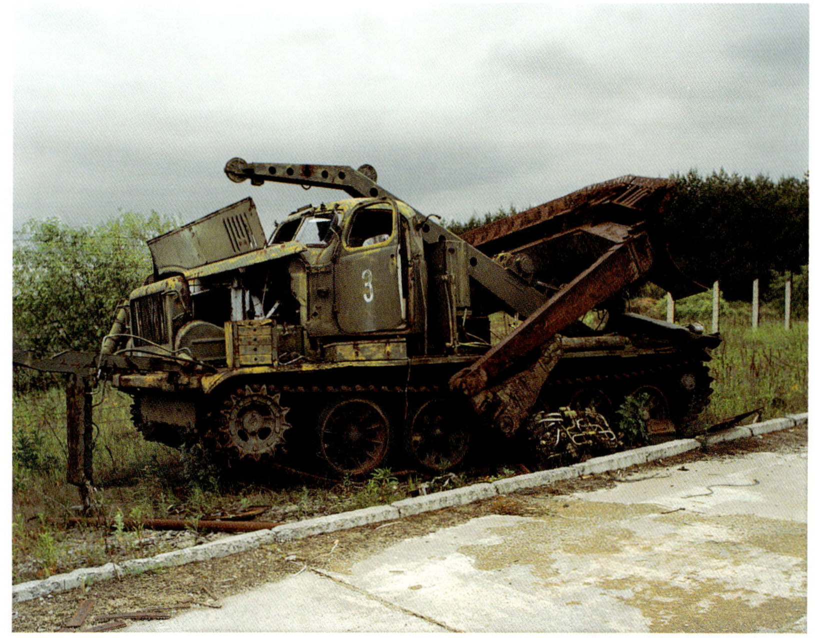
3
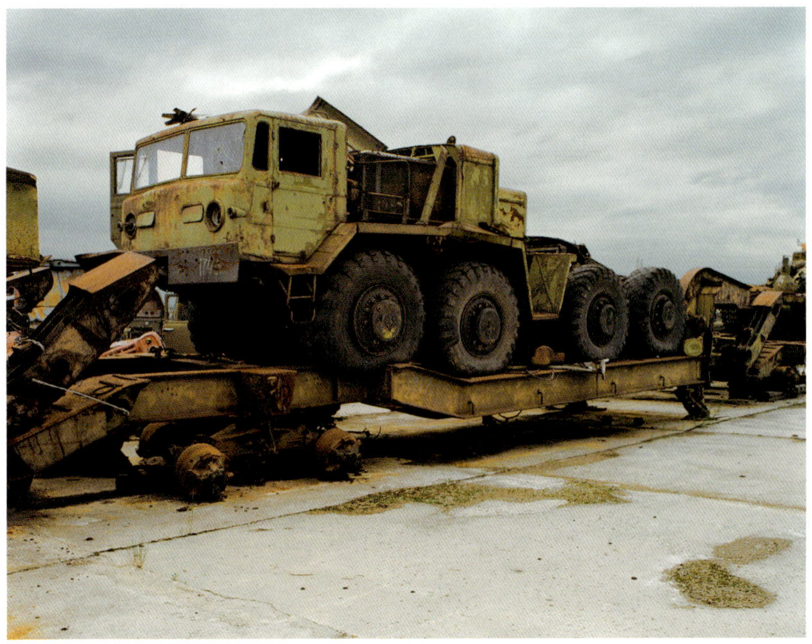
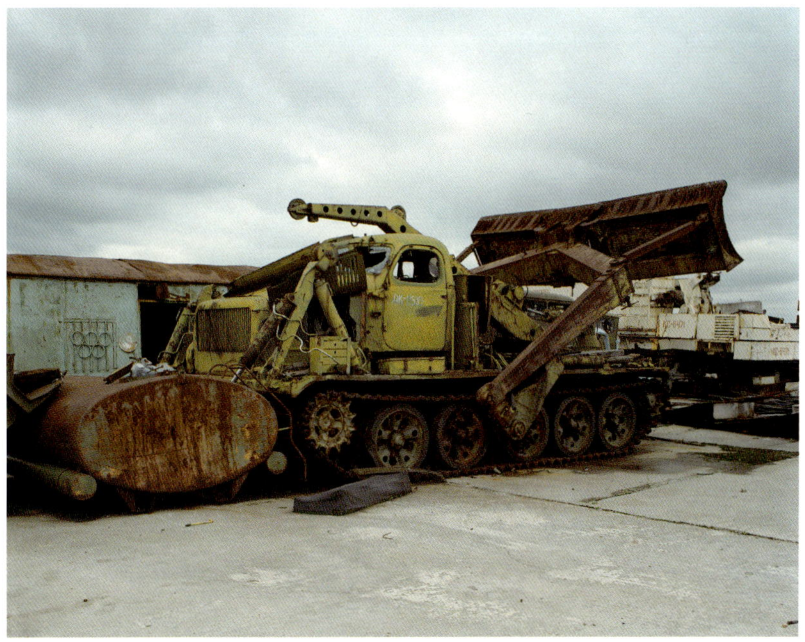
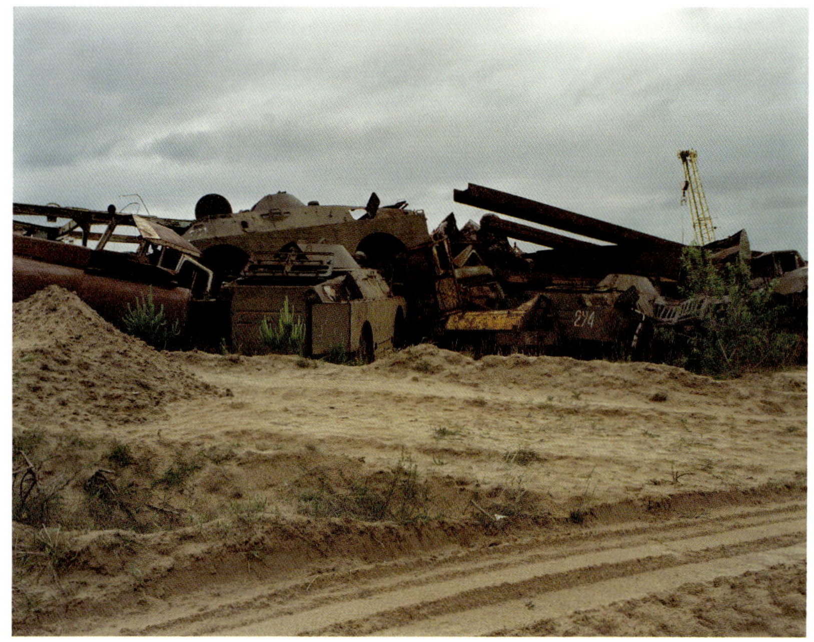

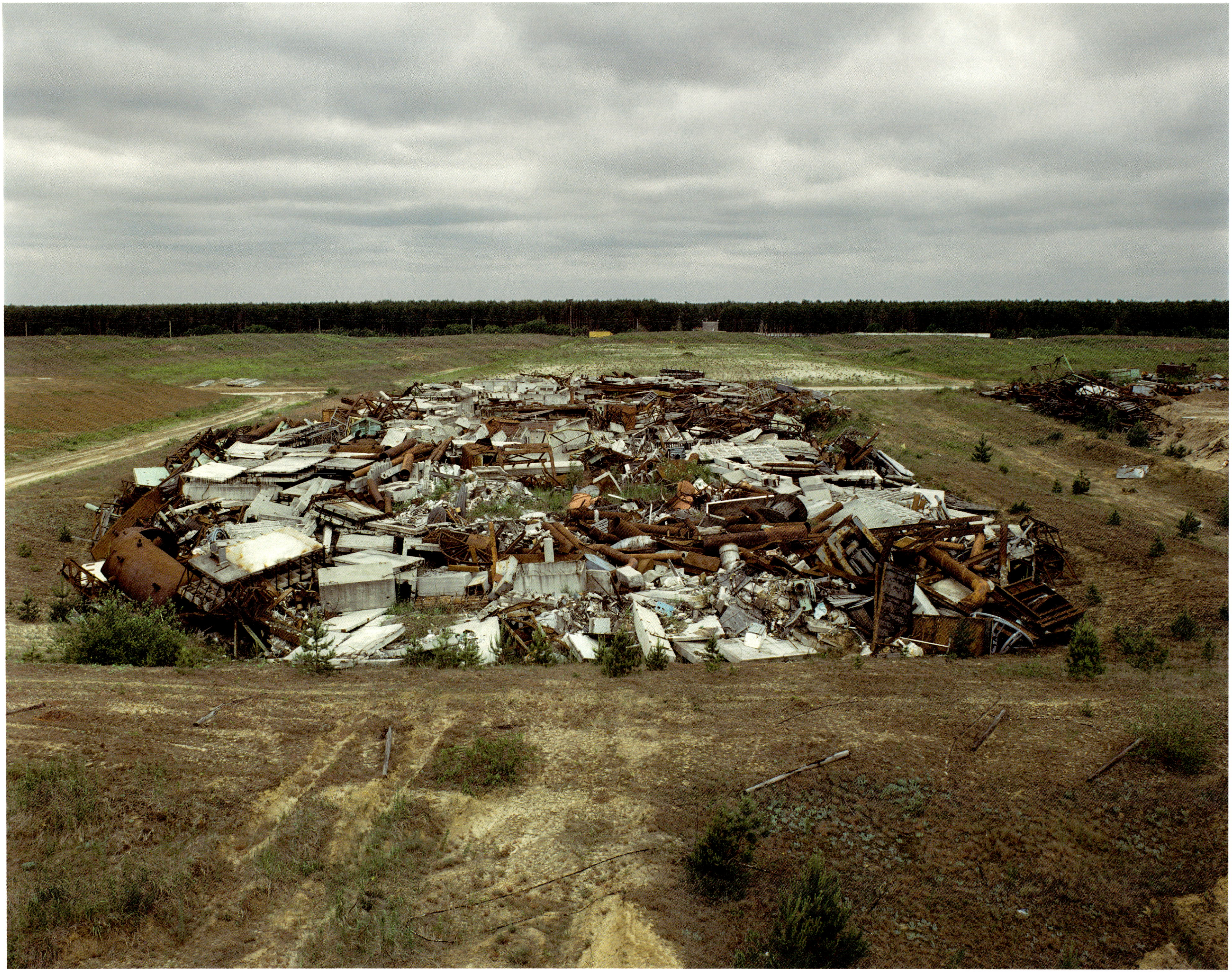

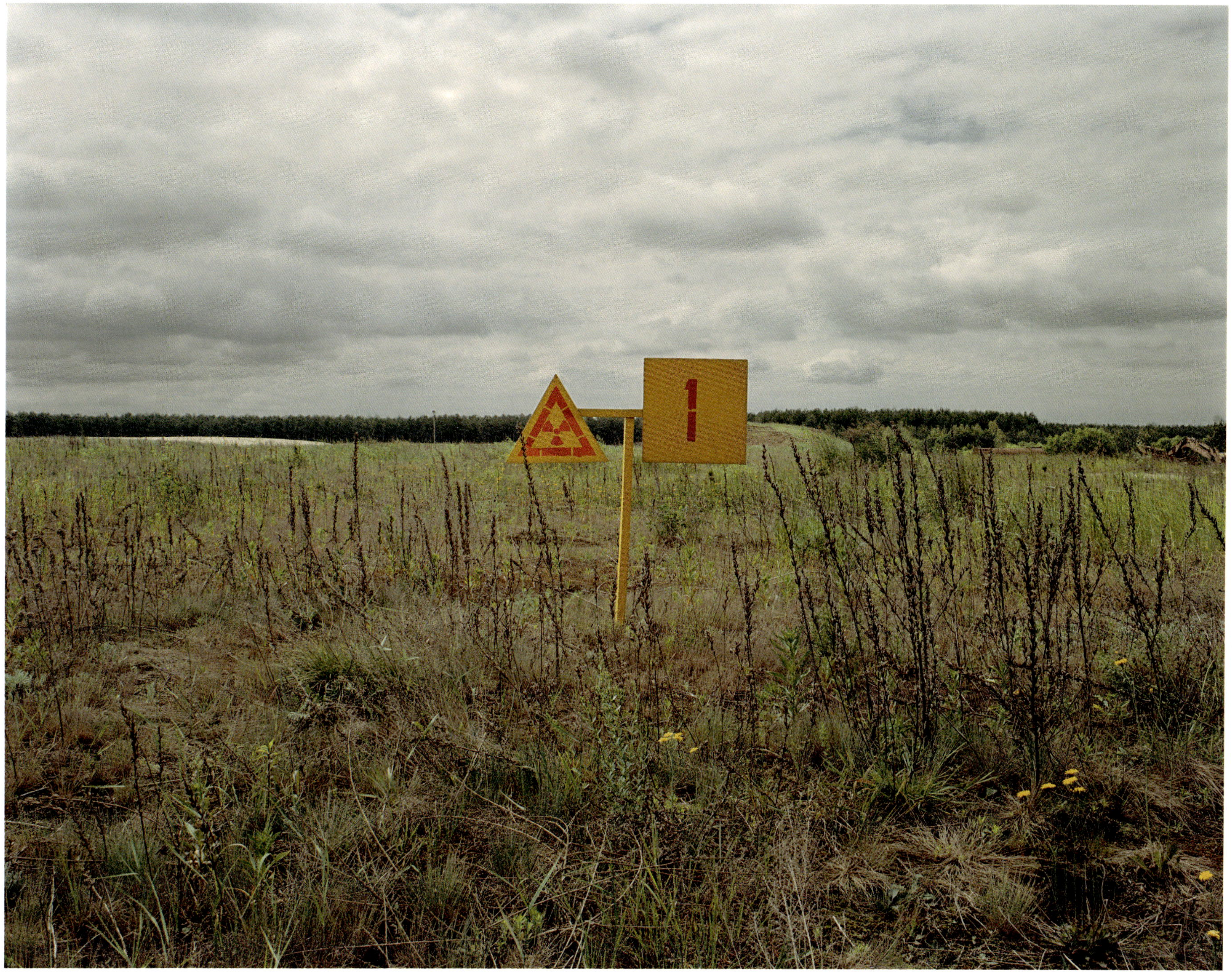
1

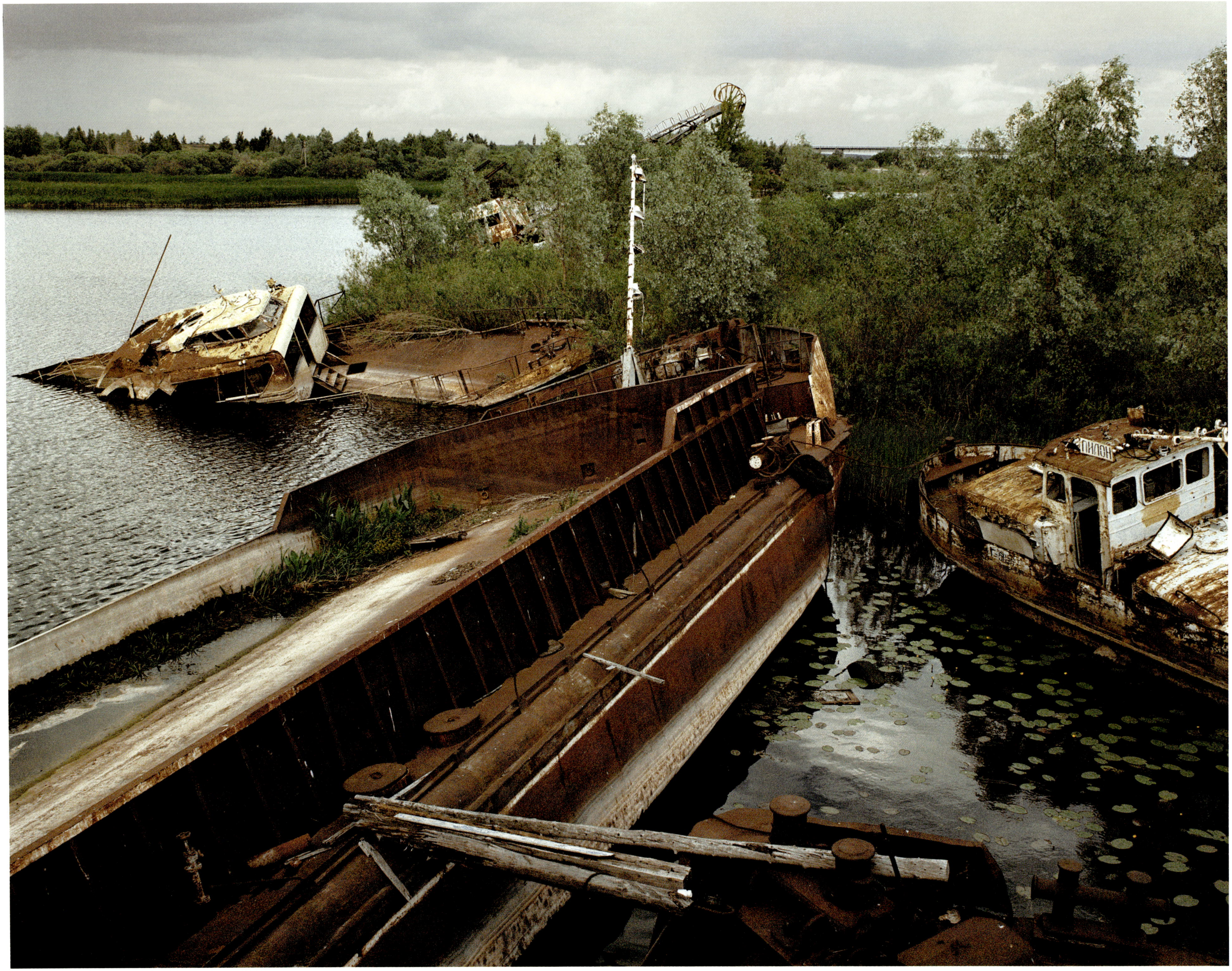
ПИЛОН

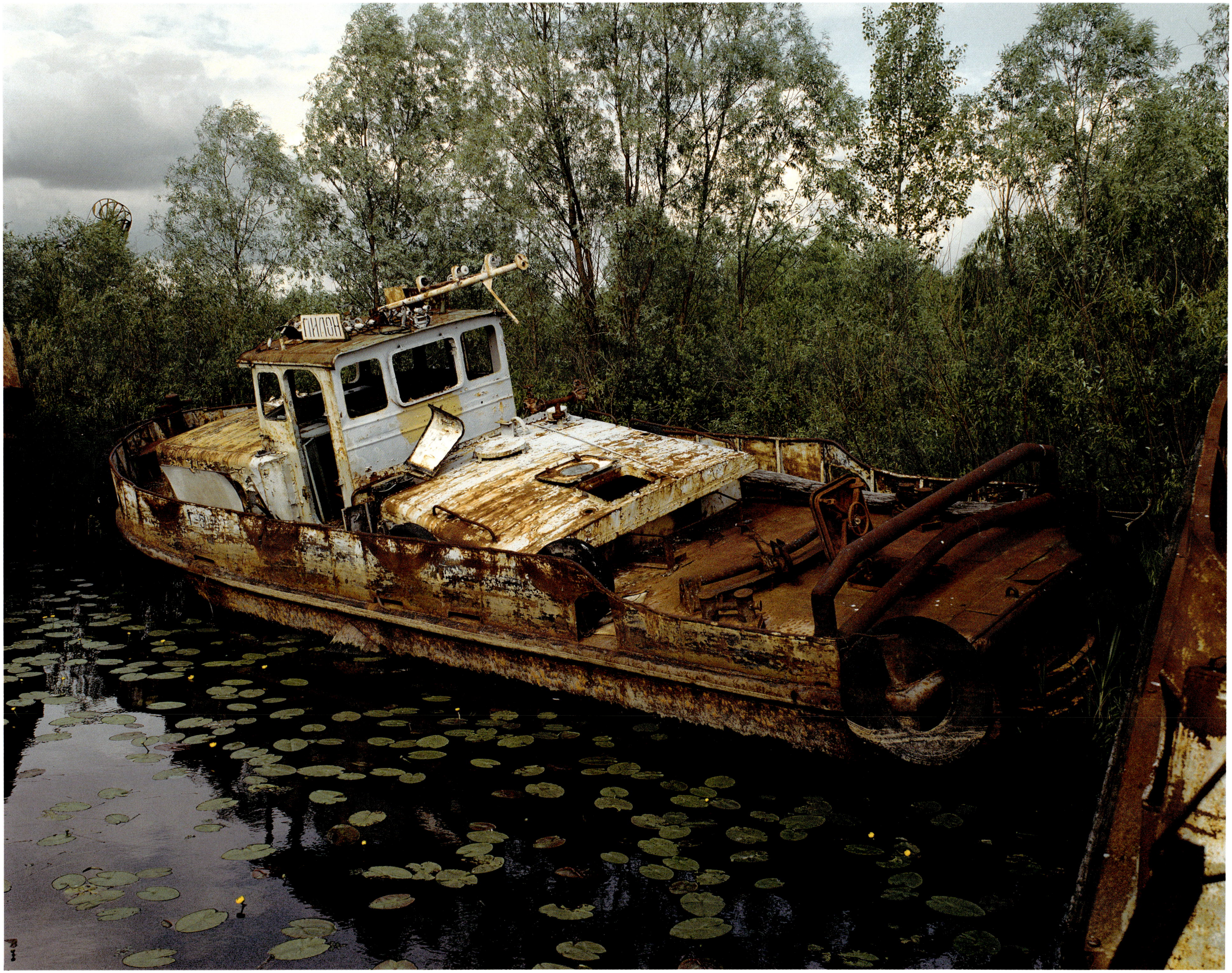

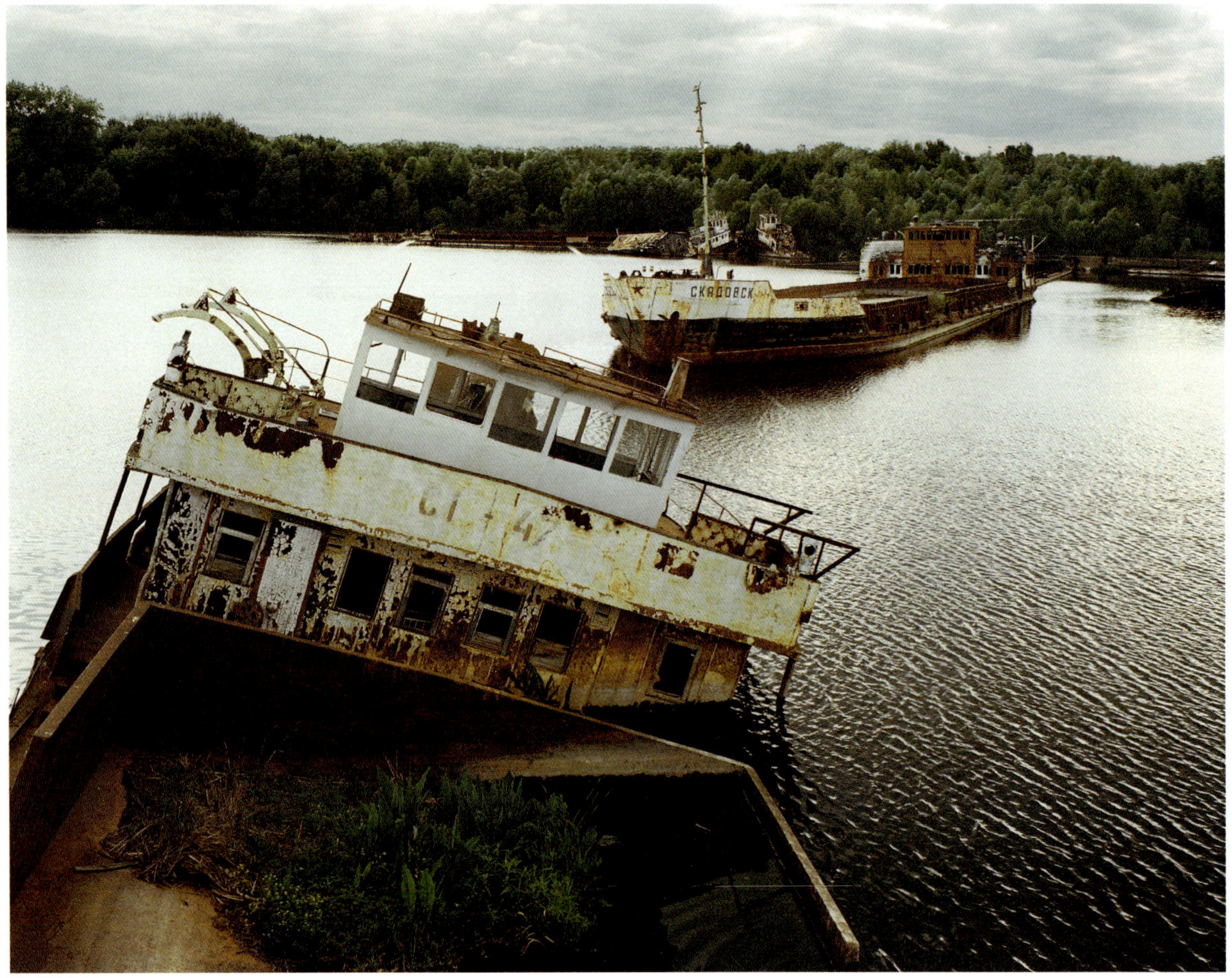
СКАДОВСК

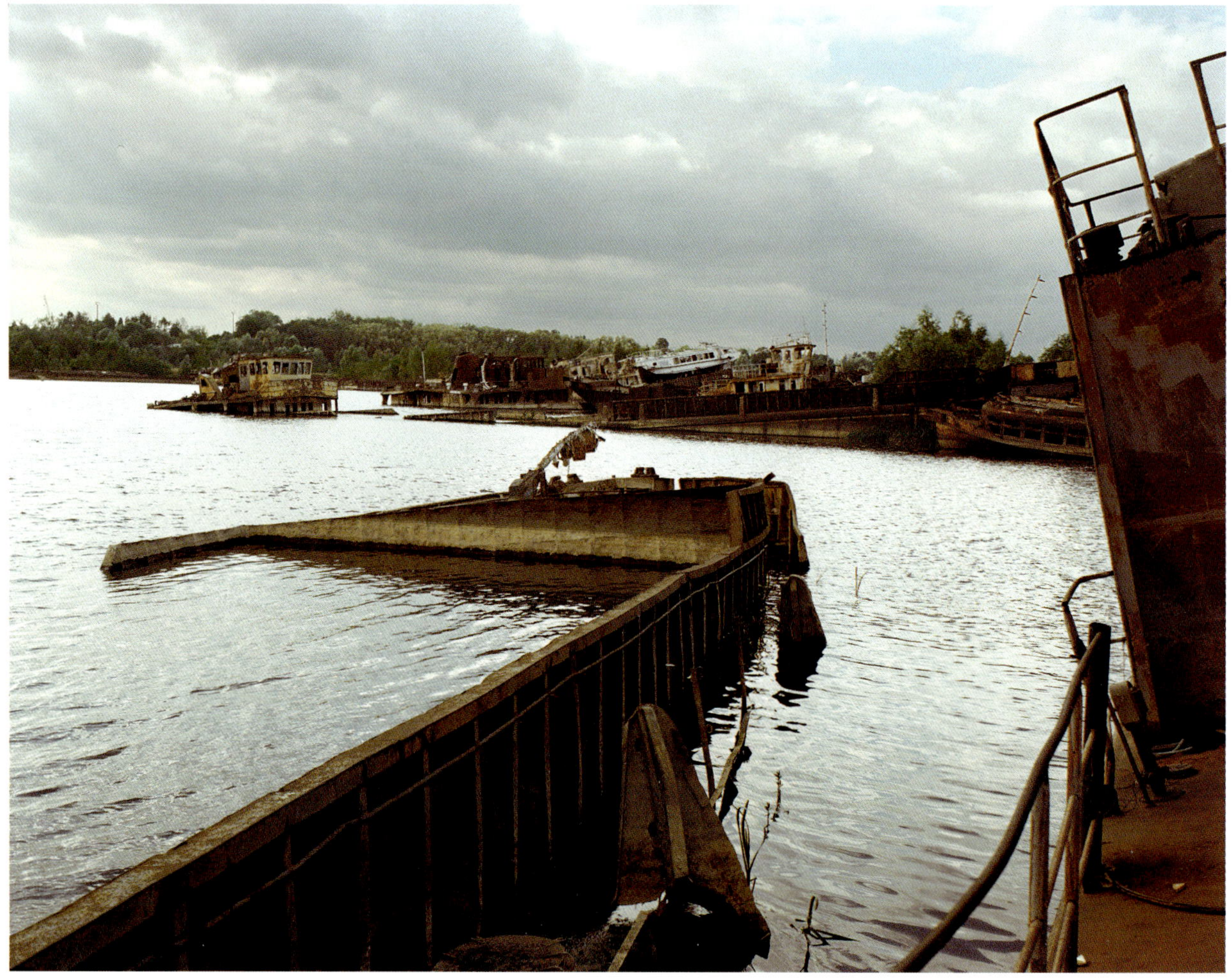

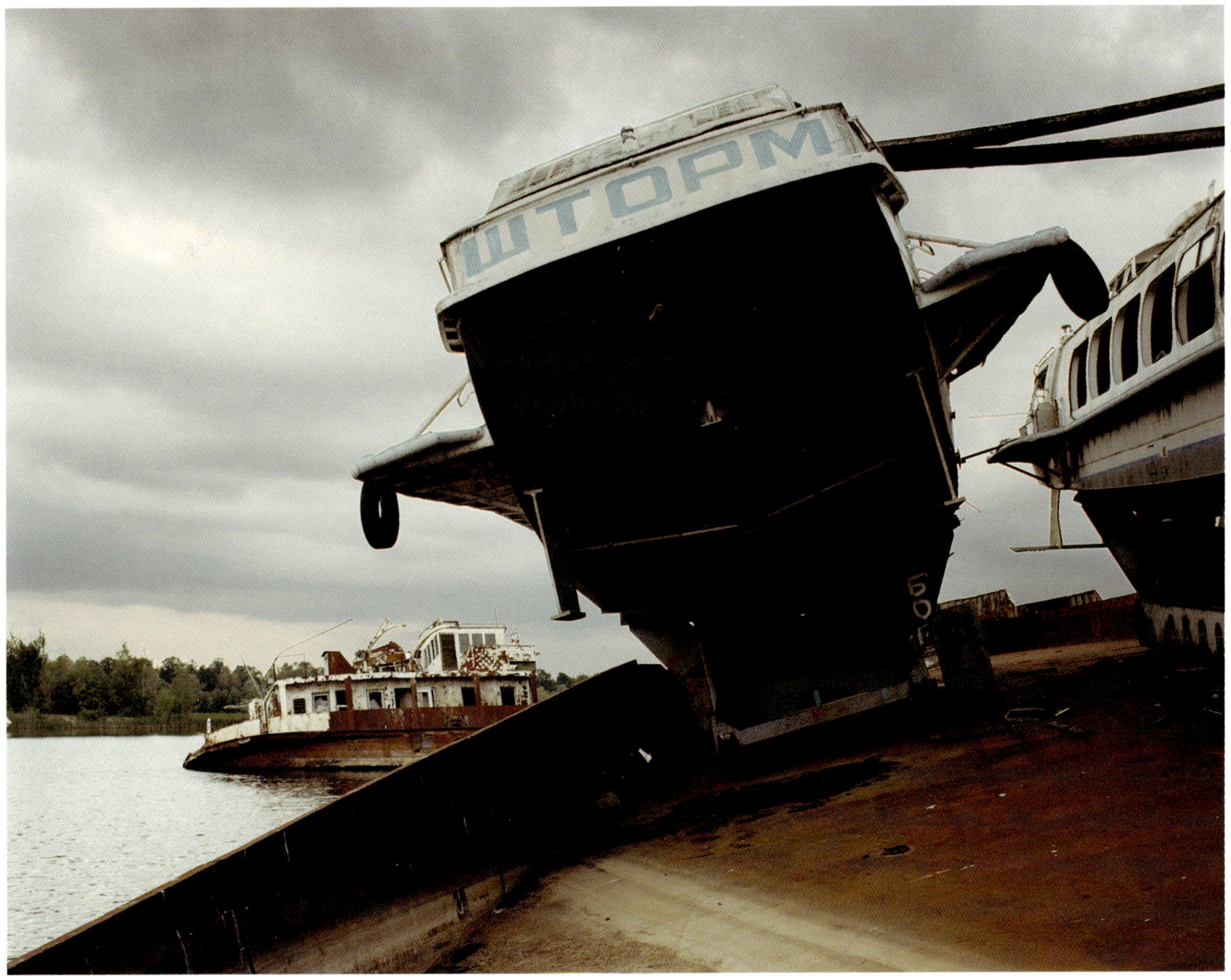
ШТОРМ

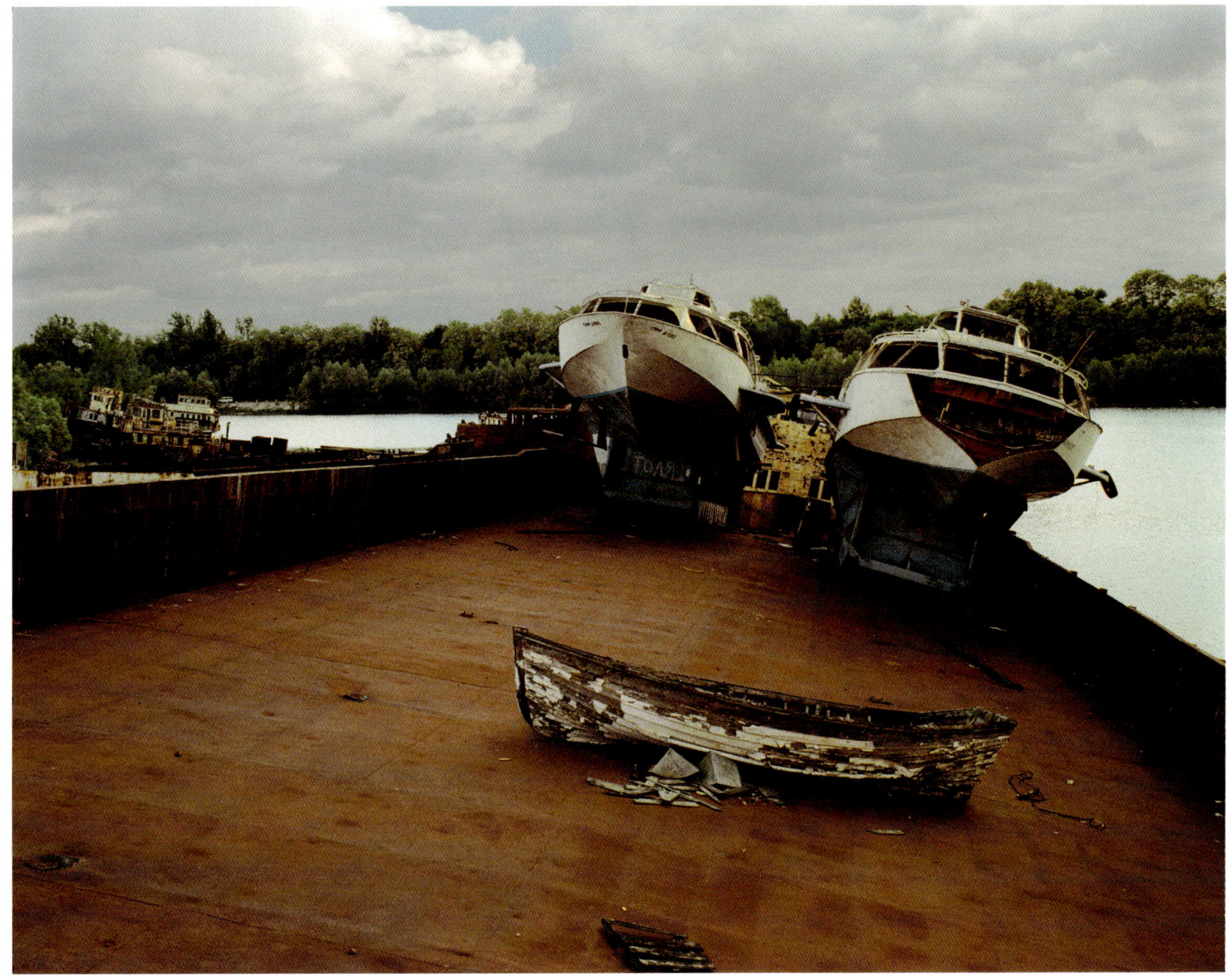

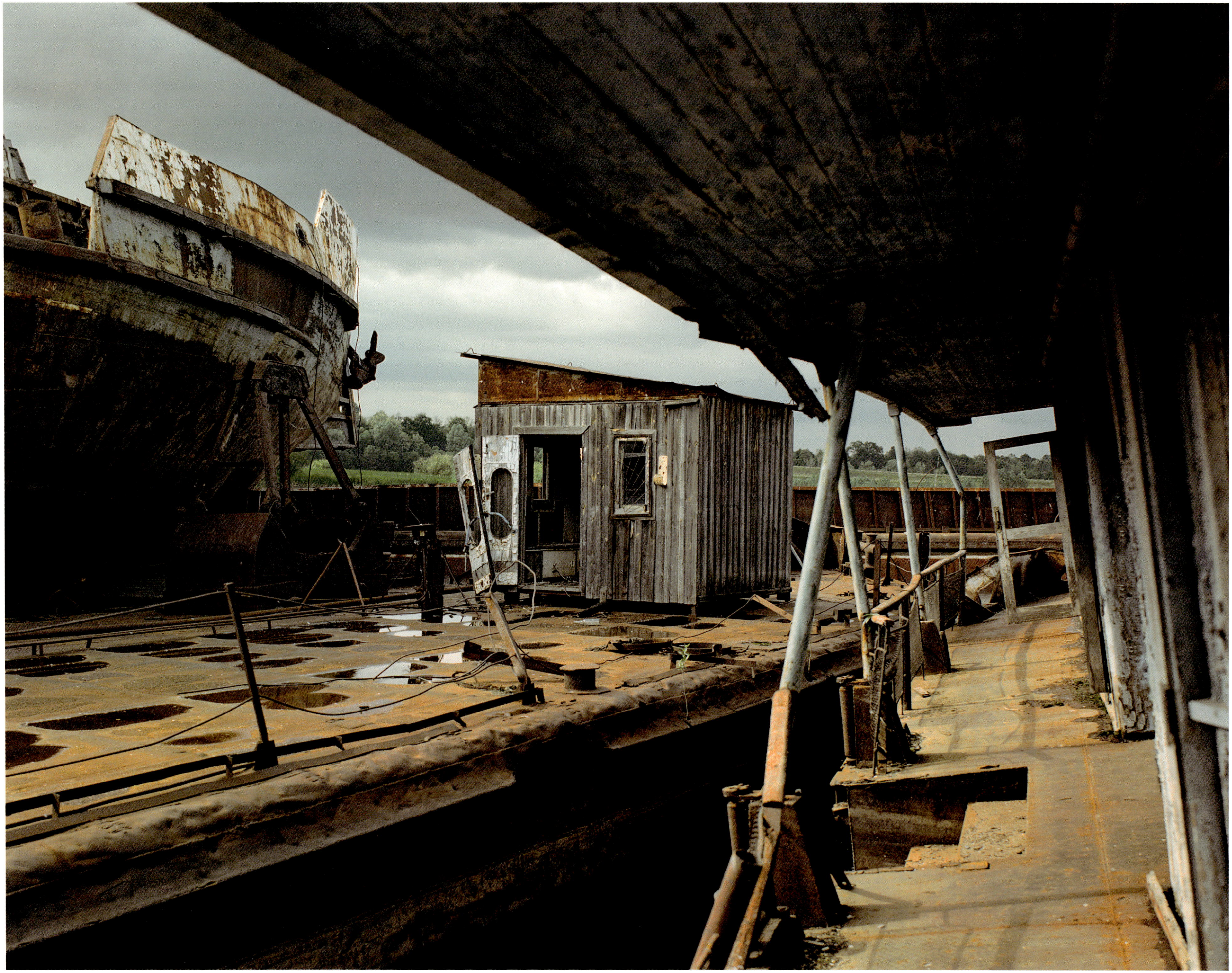

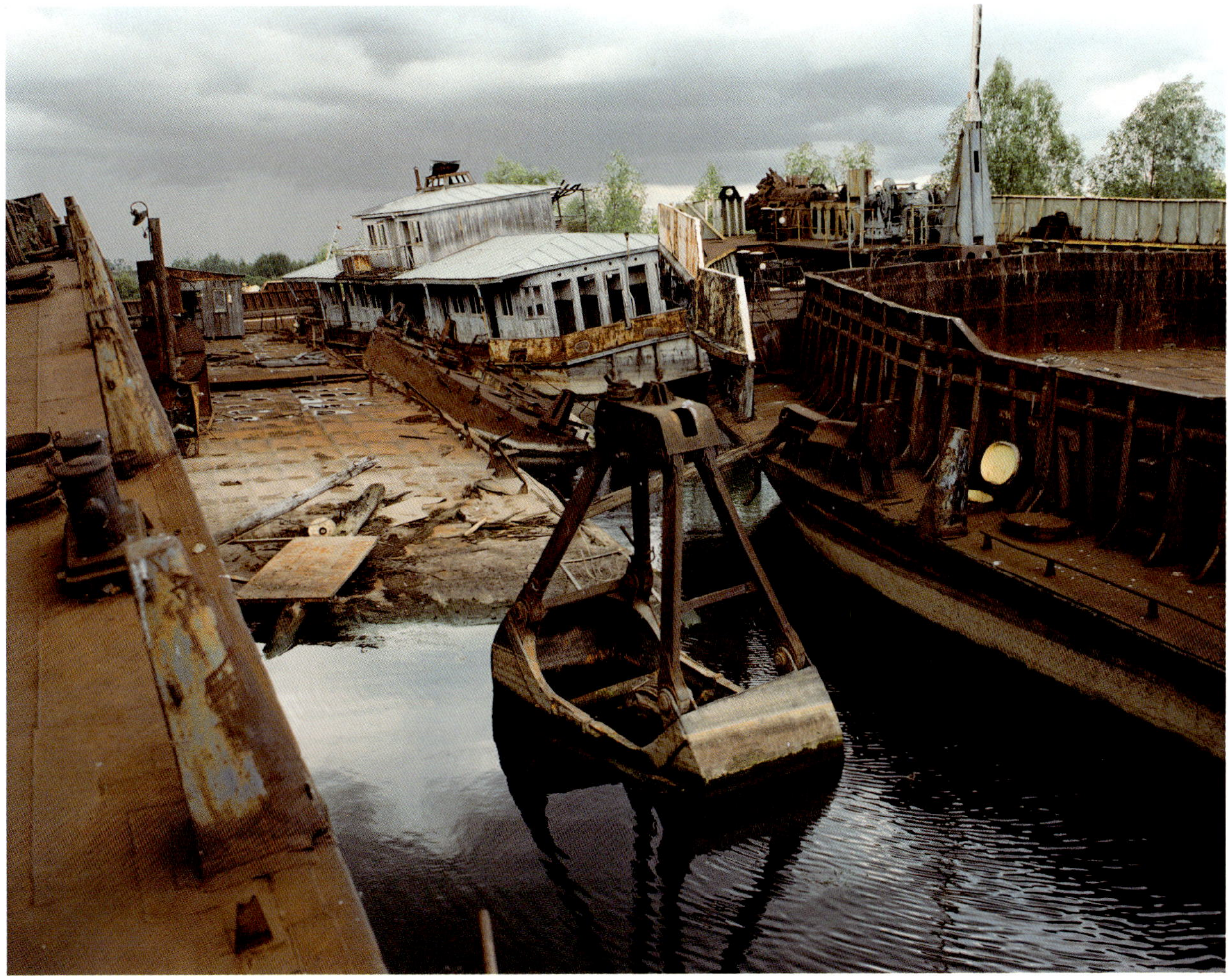

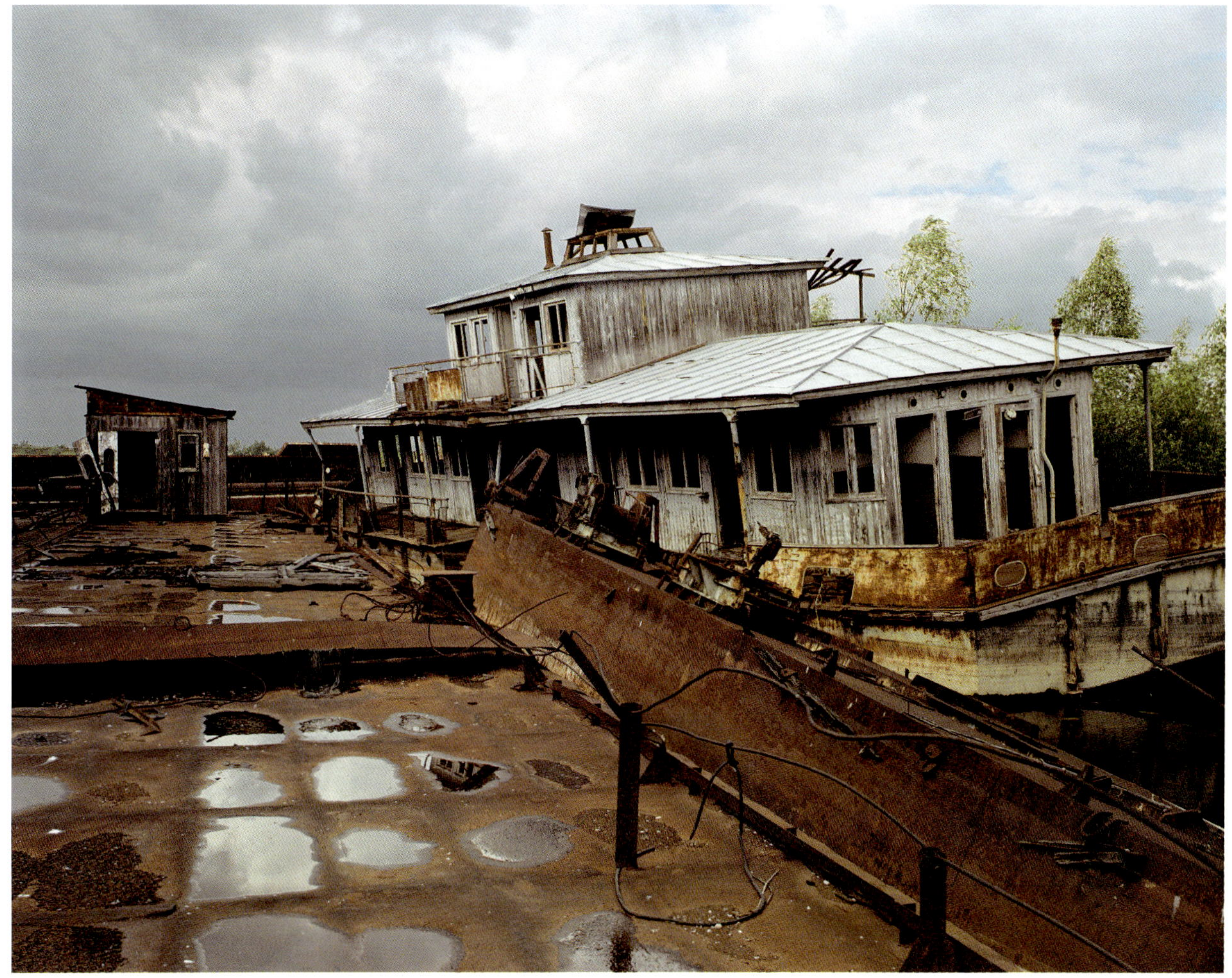

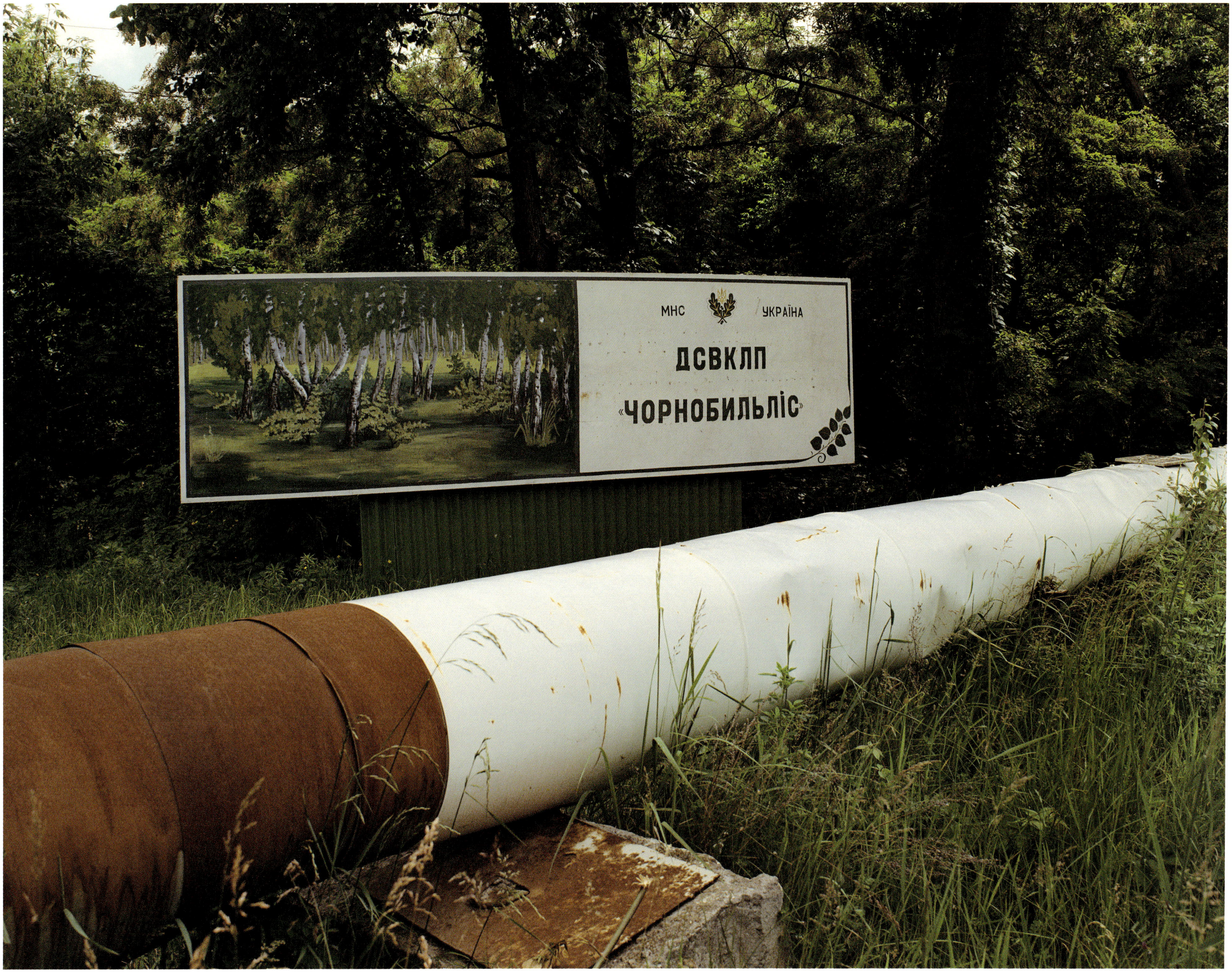
МНС
УКРАЇНА
ДСВКЛП
«ЧОРНОБИЛЬЛІС»

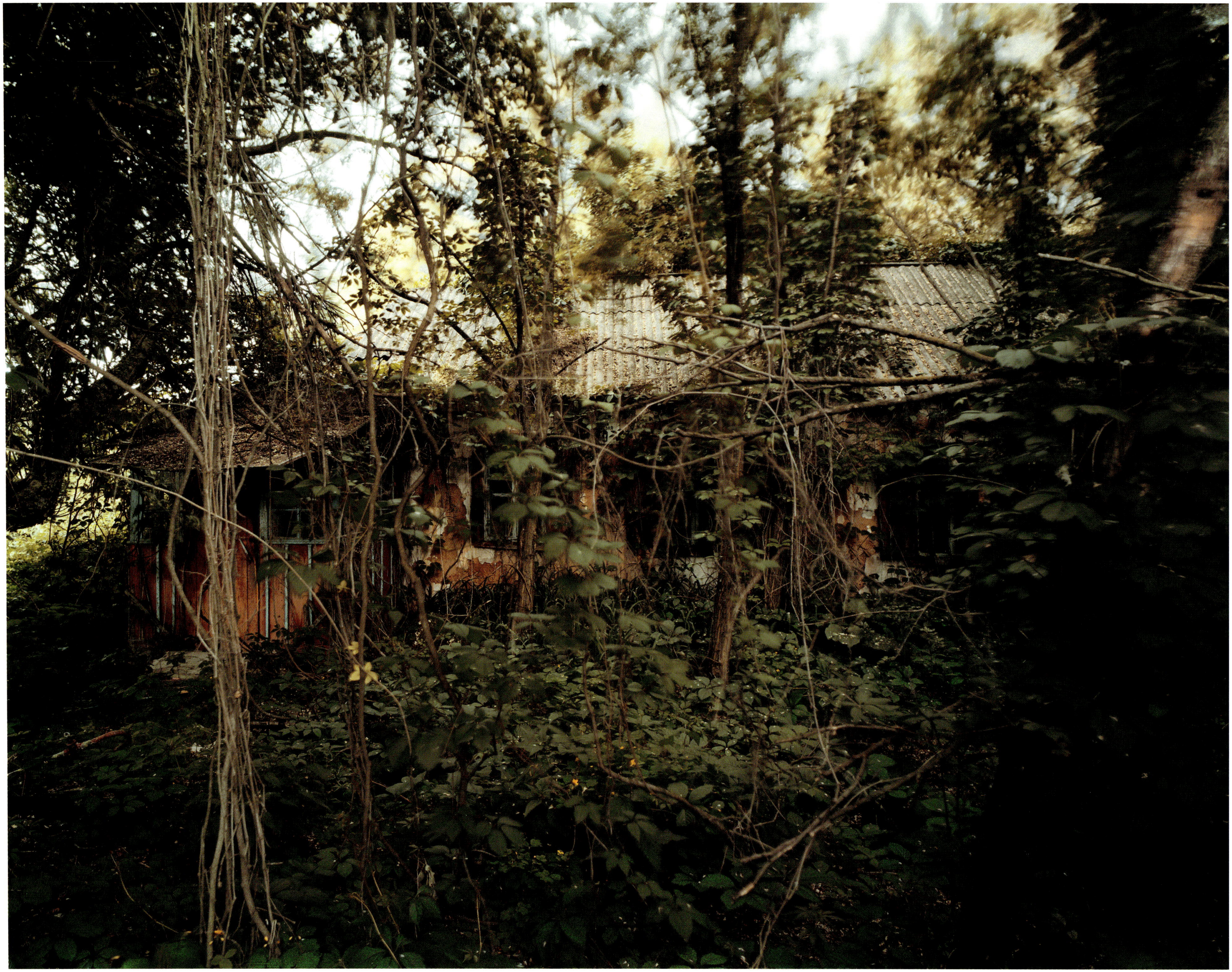

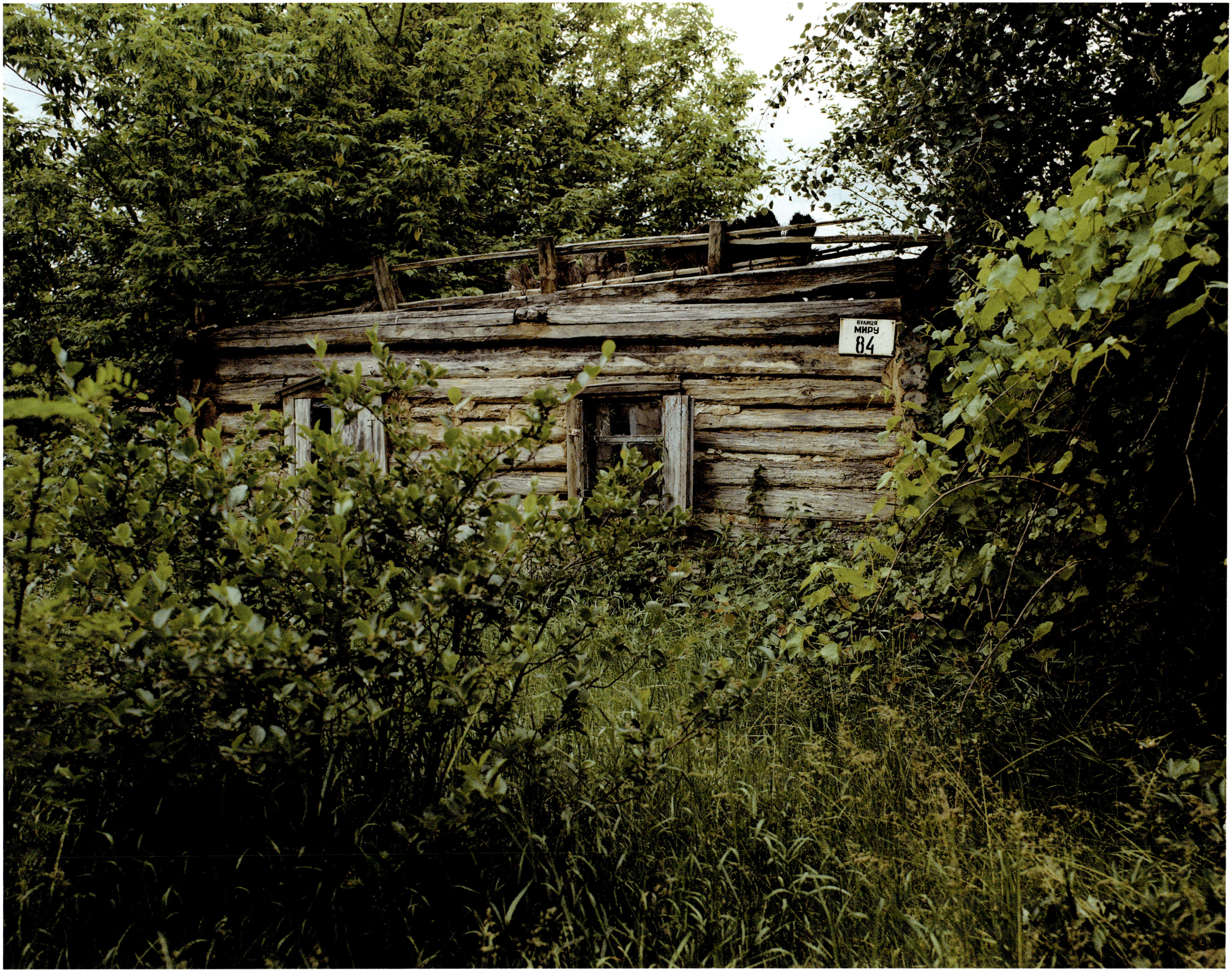
ВУЛИЦЯ
МИРУ
84

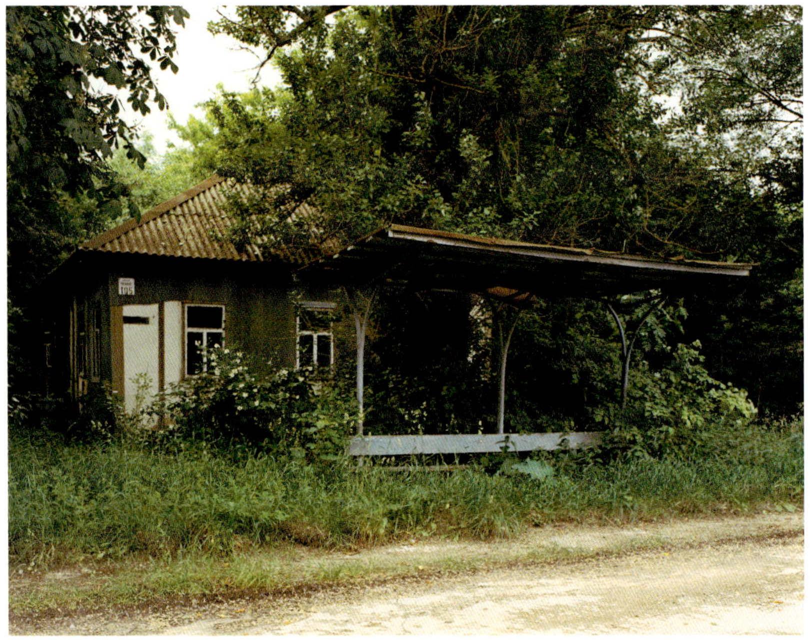
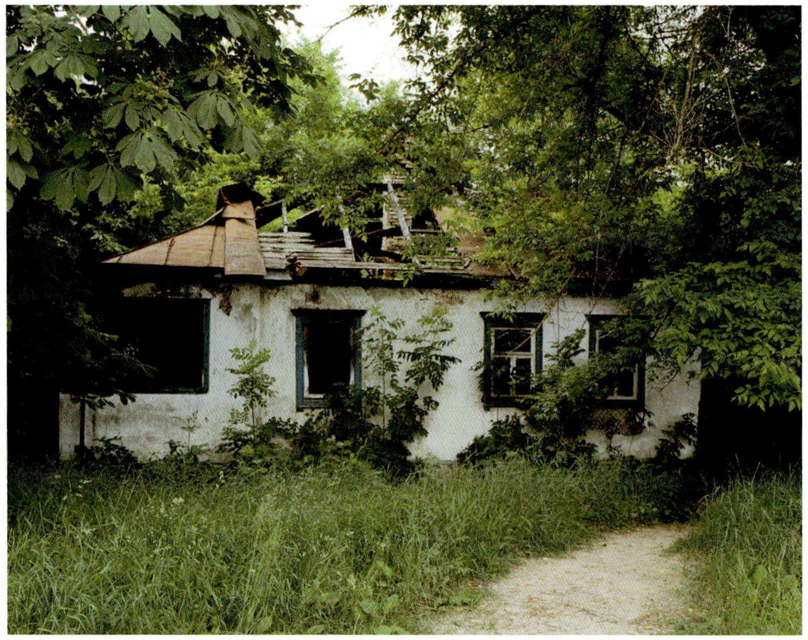
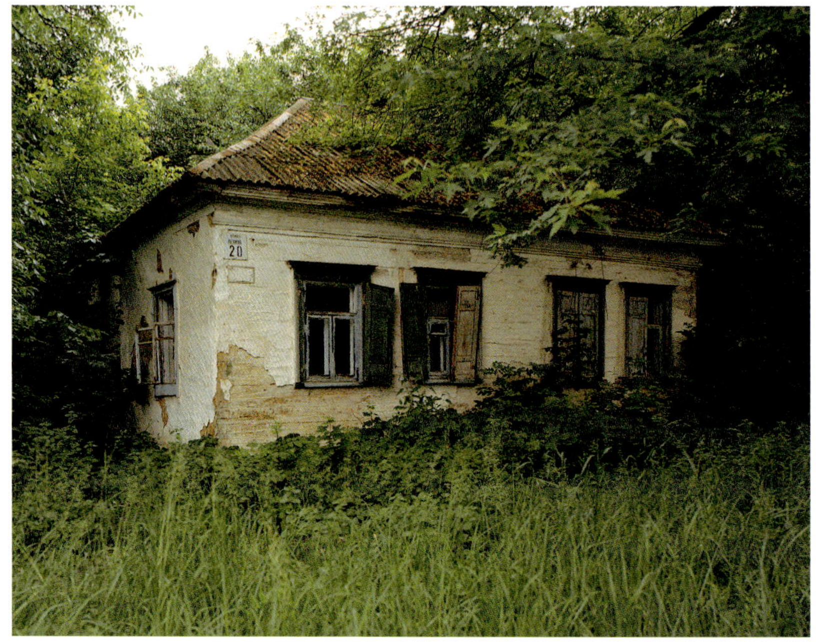
20
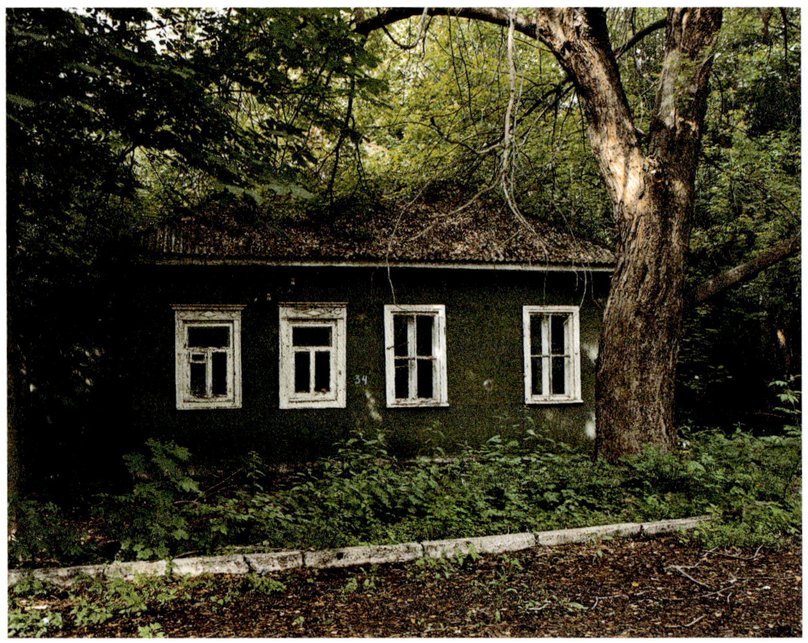
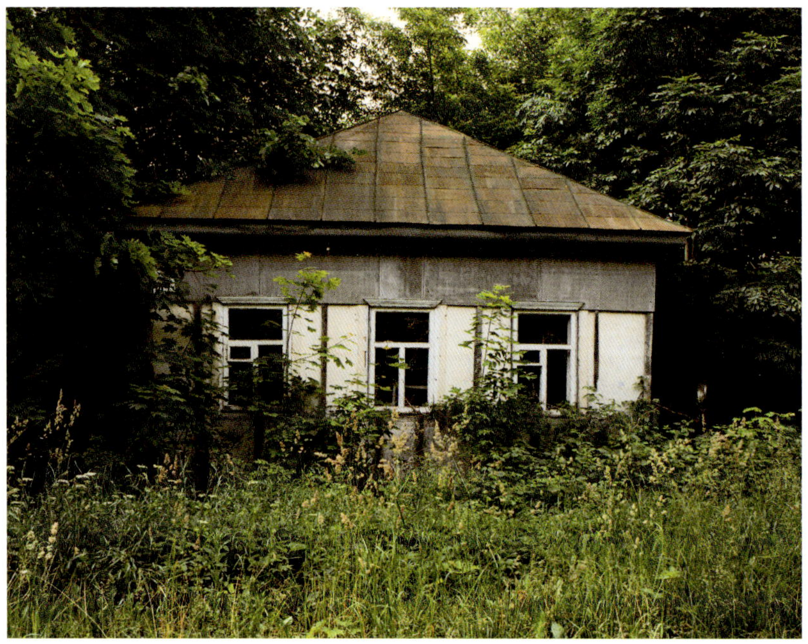
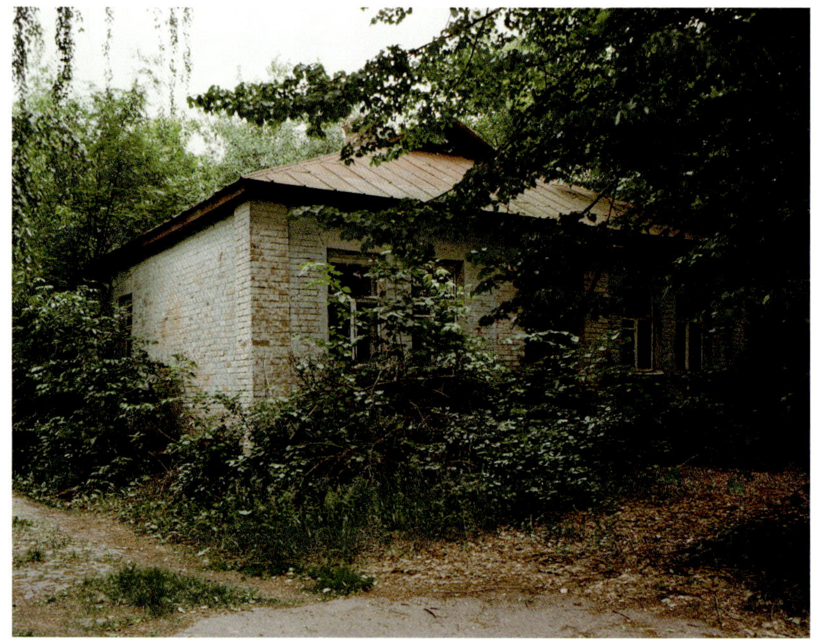

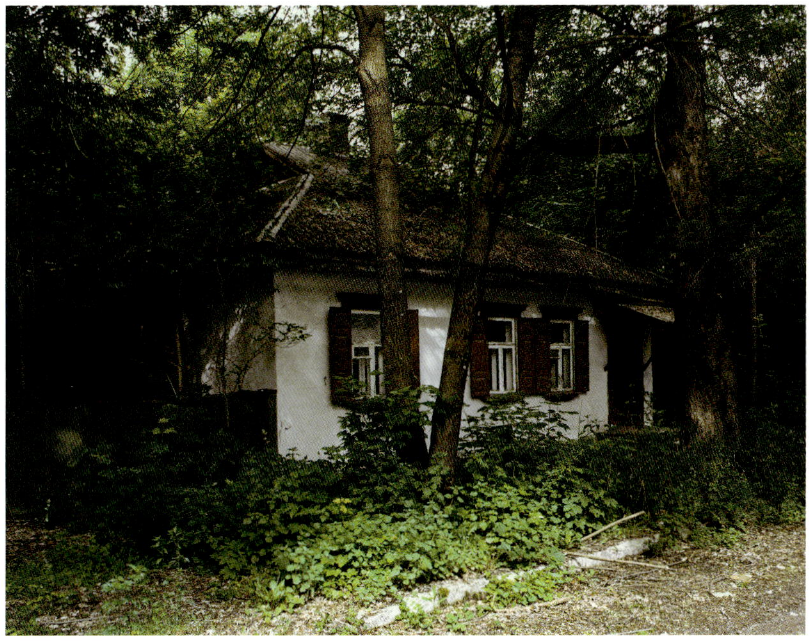
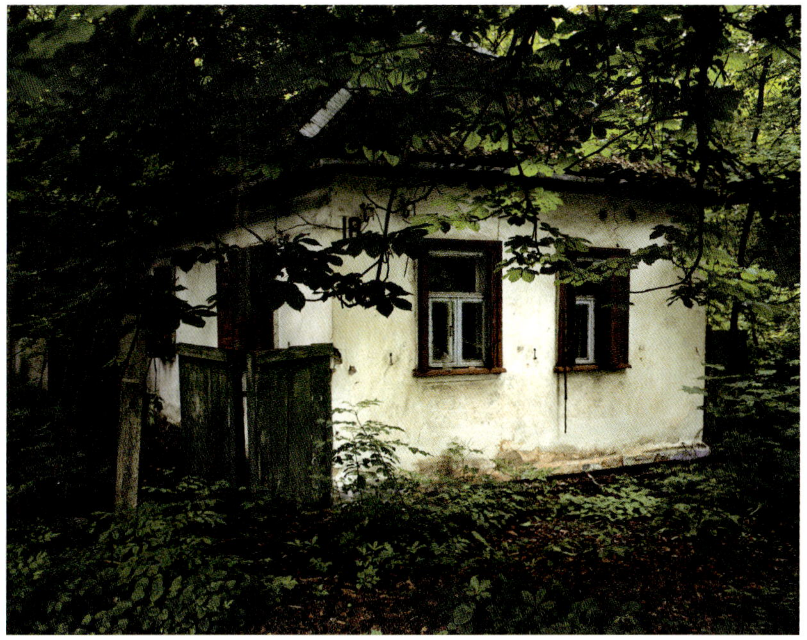
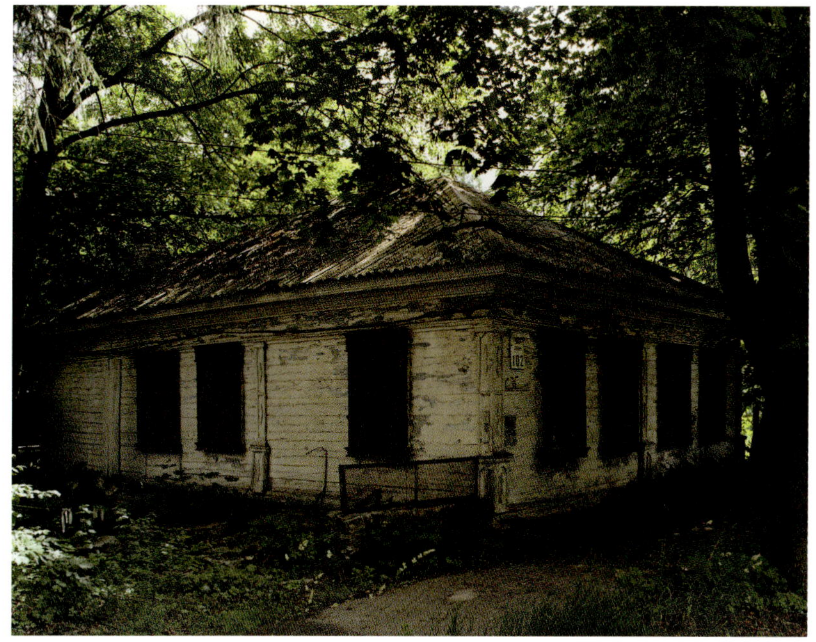
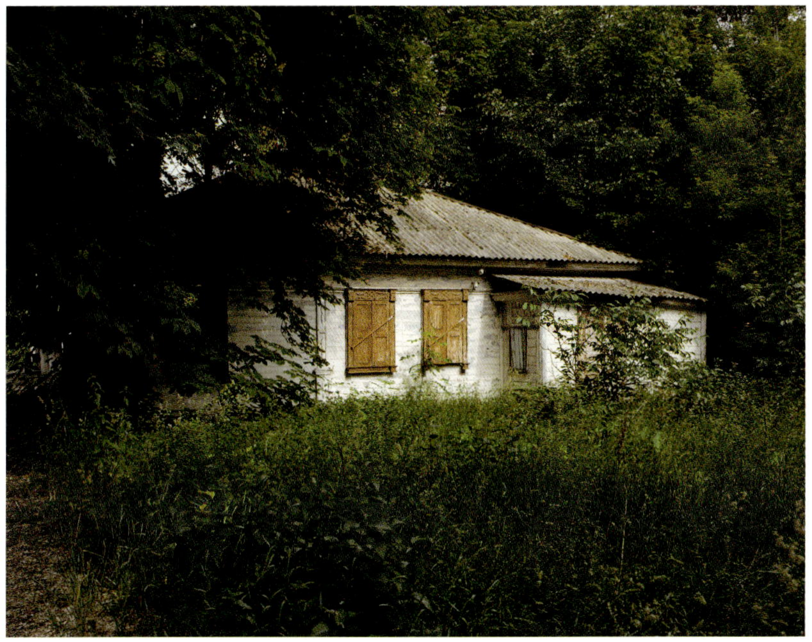
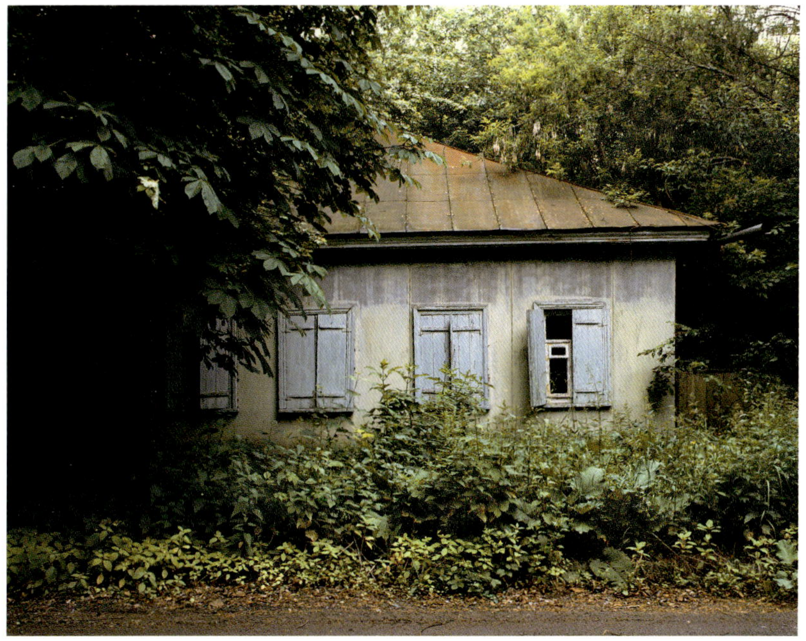
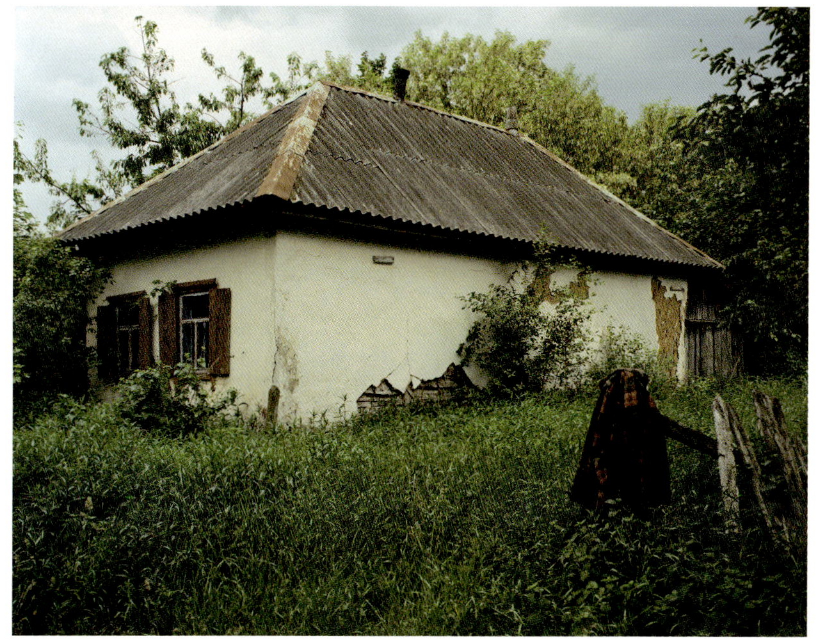

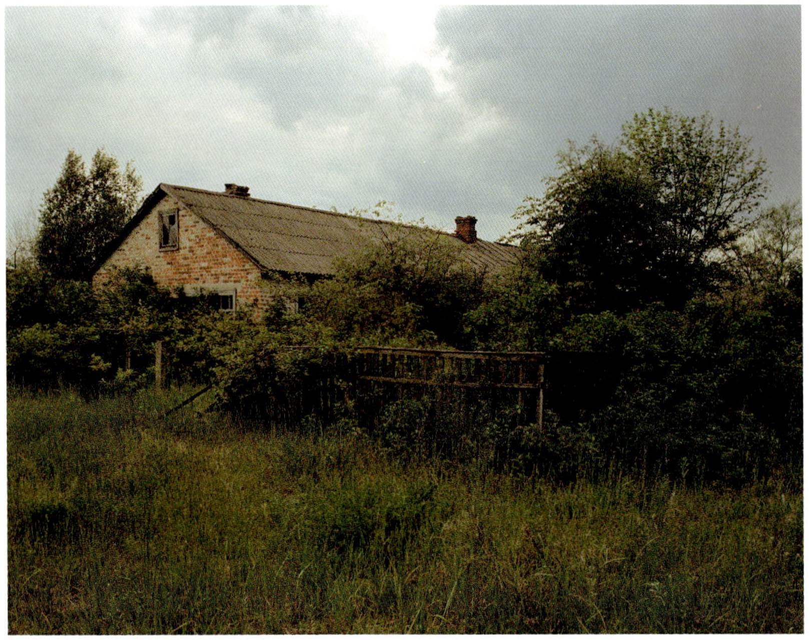
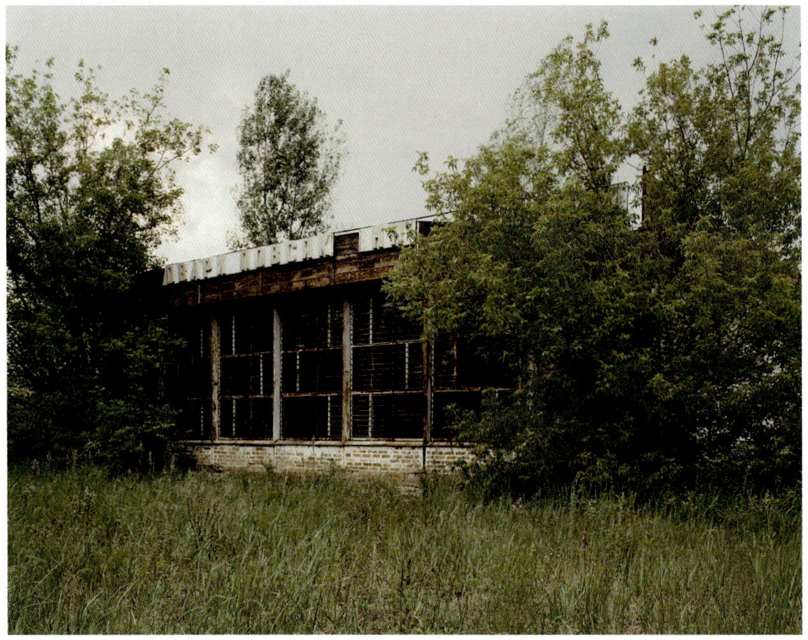
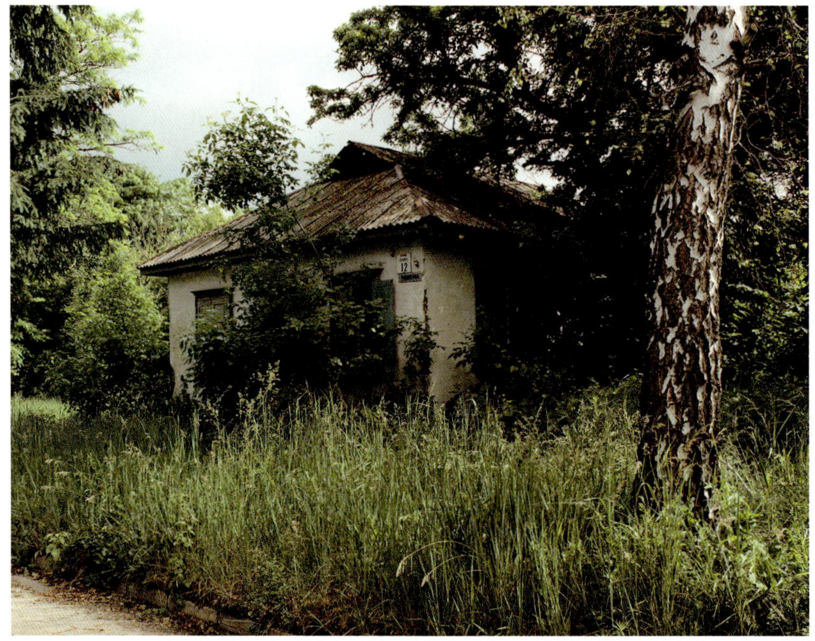
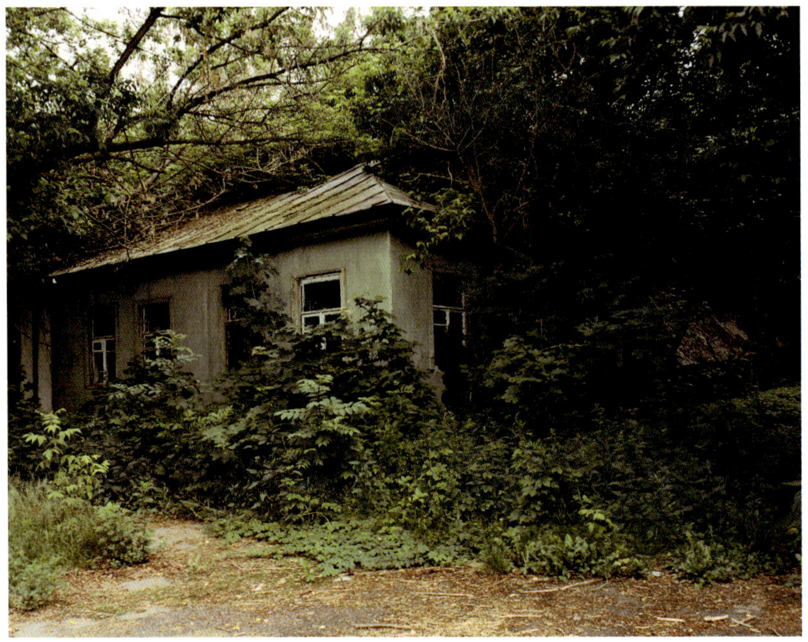
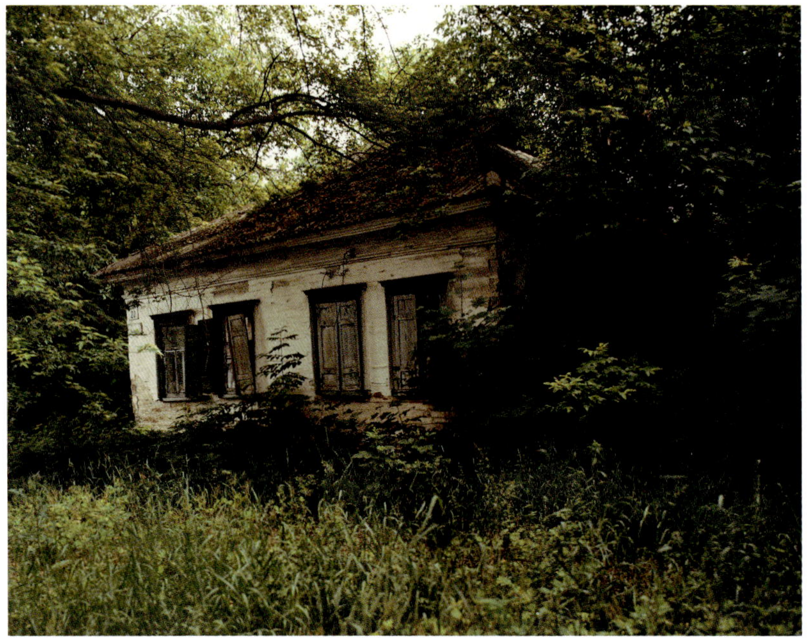
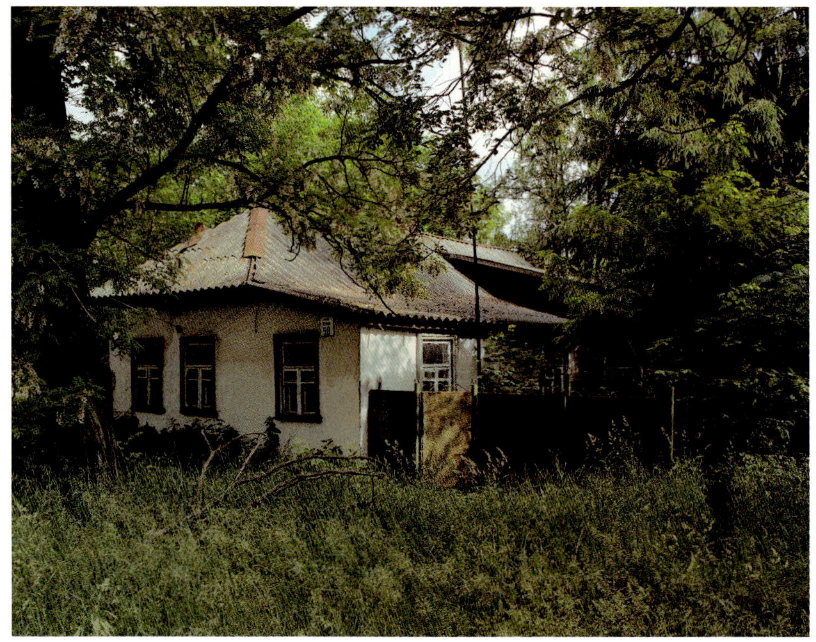

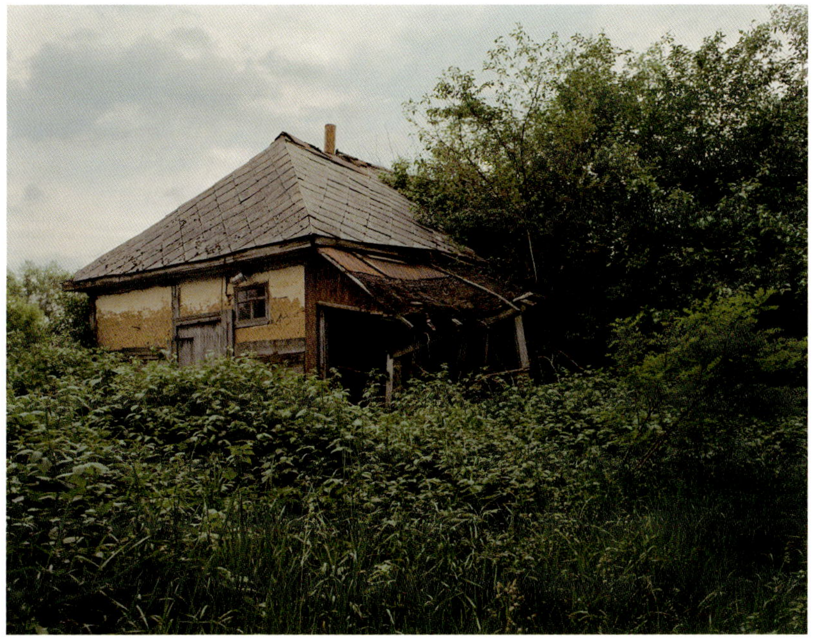
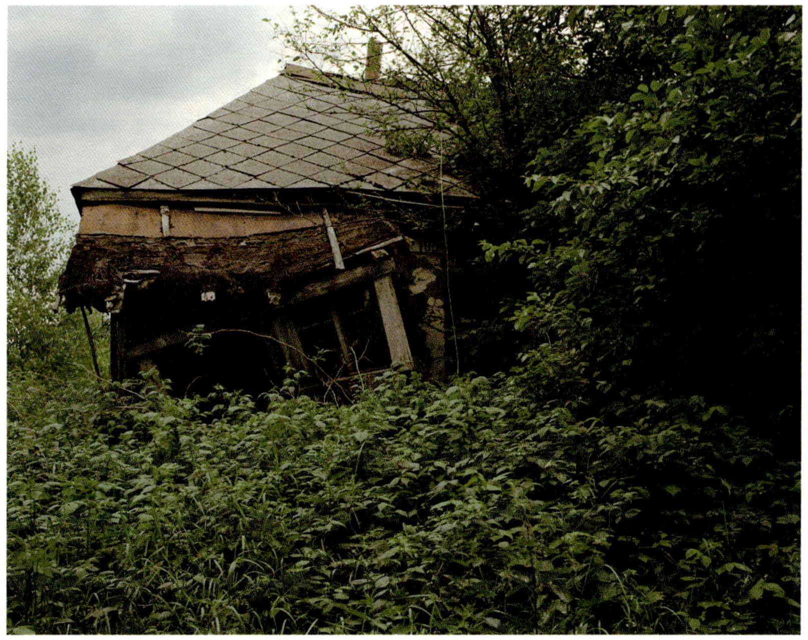
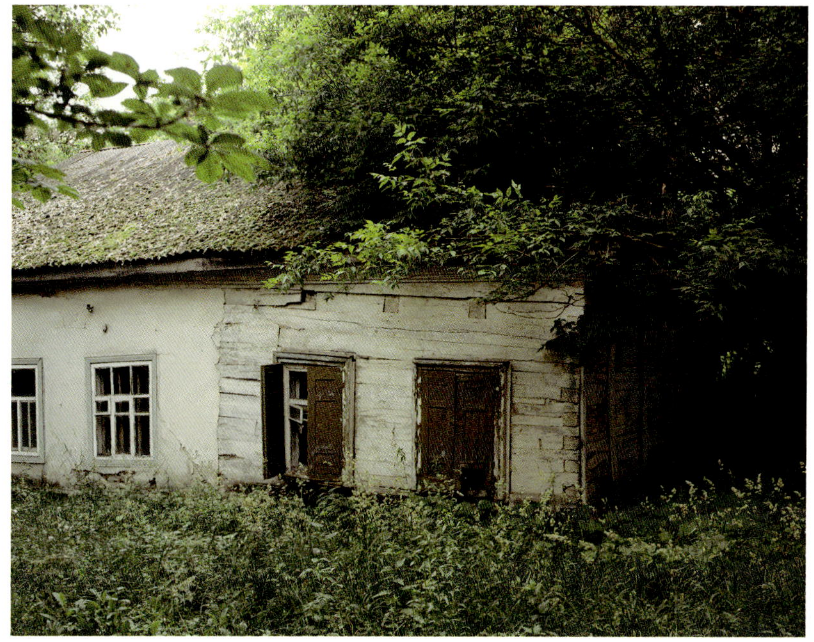
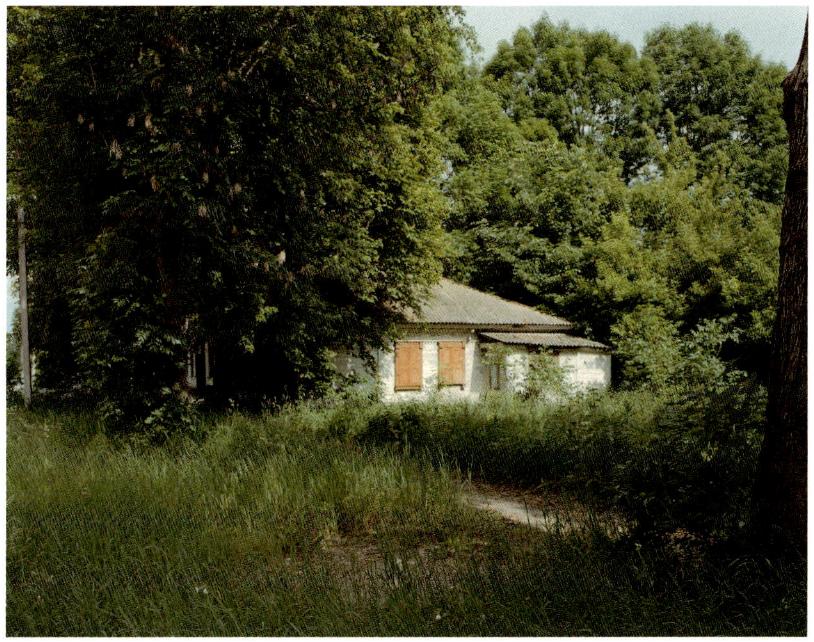
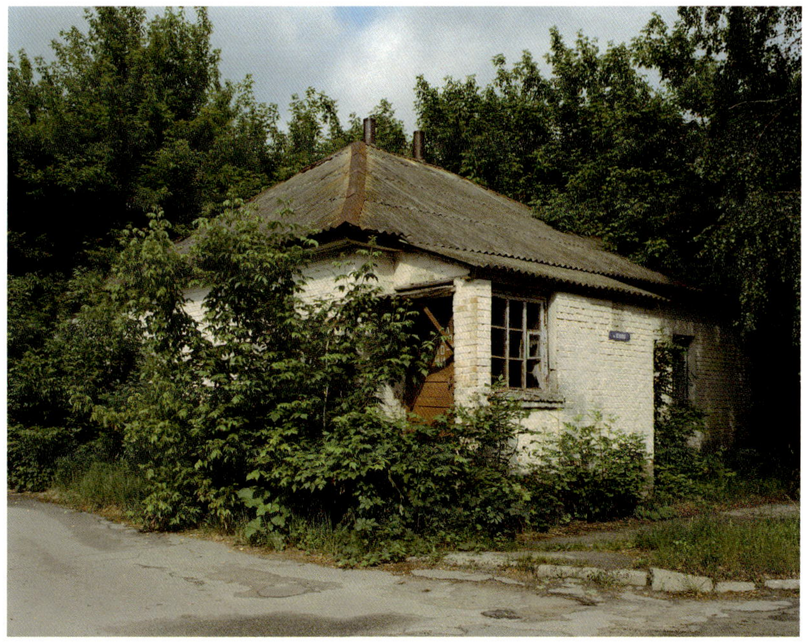
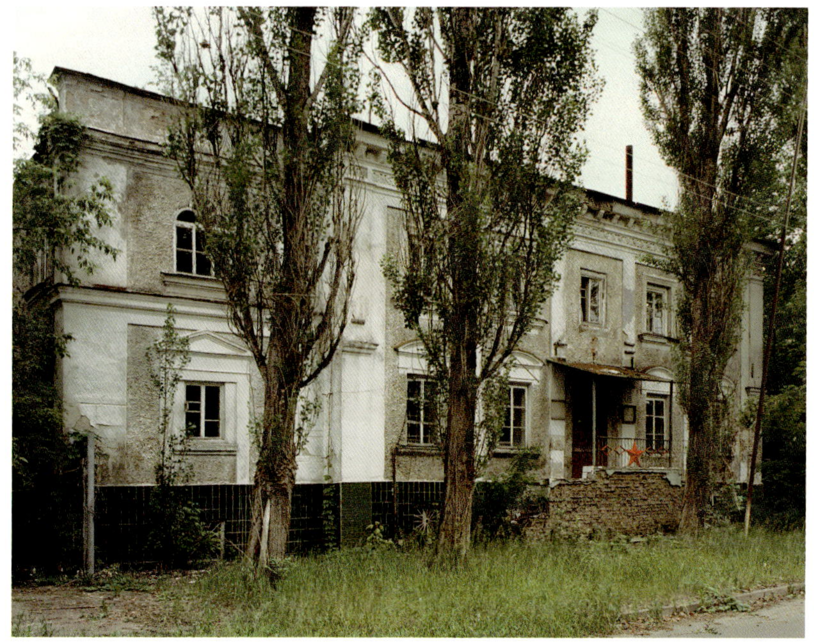

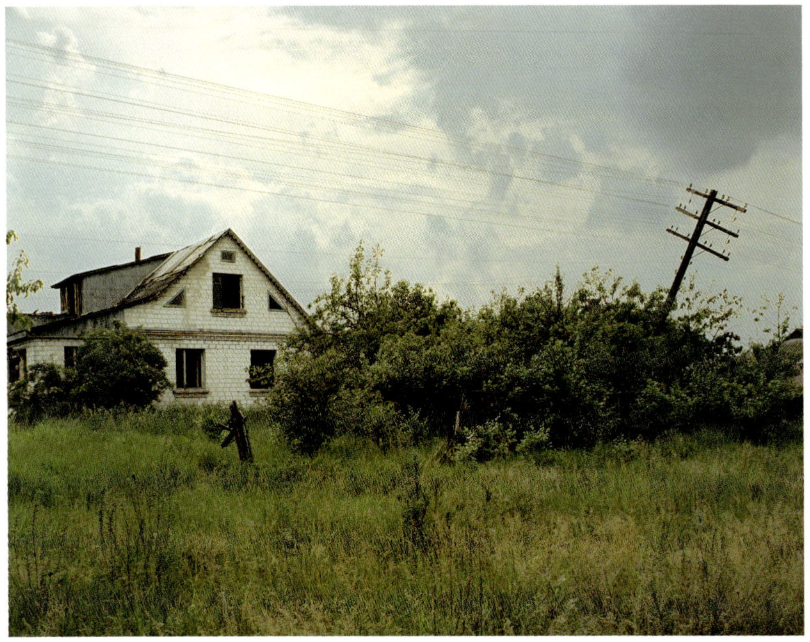
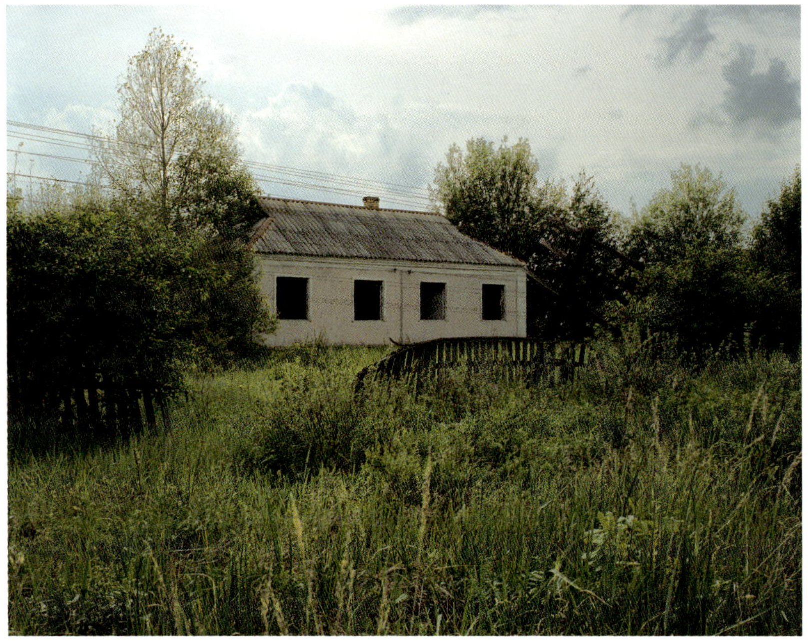
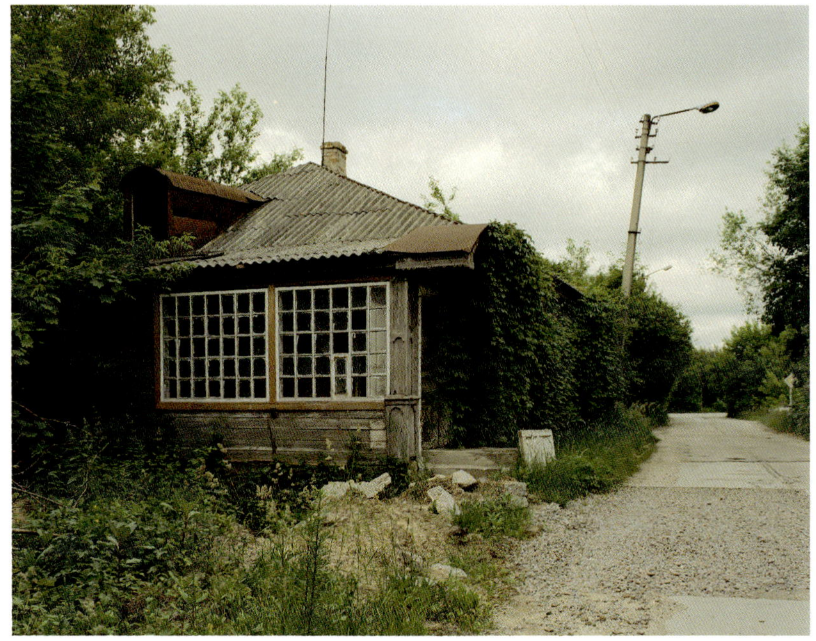
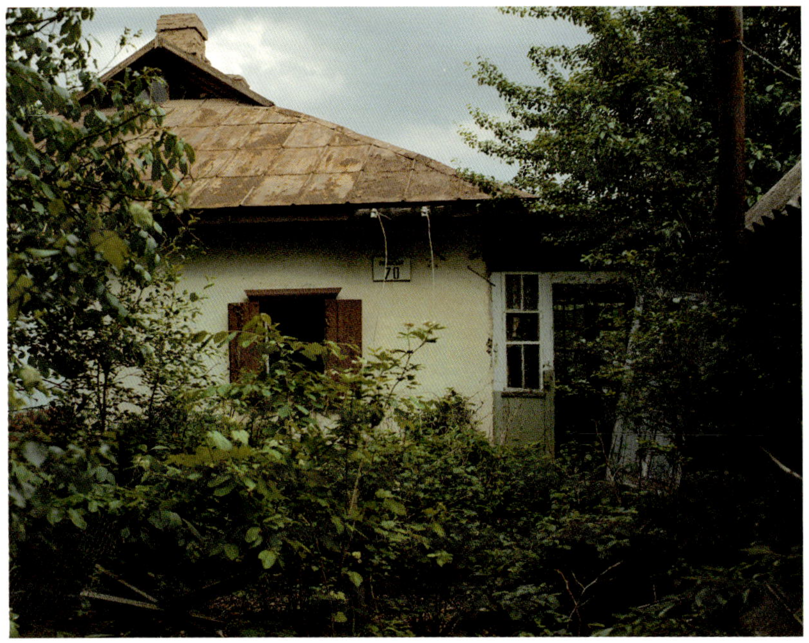
70
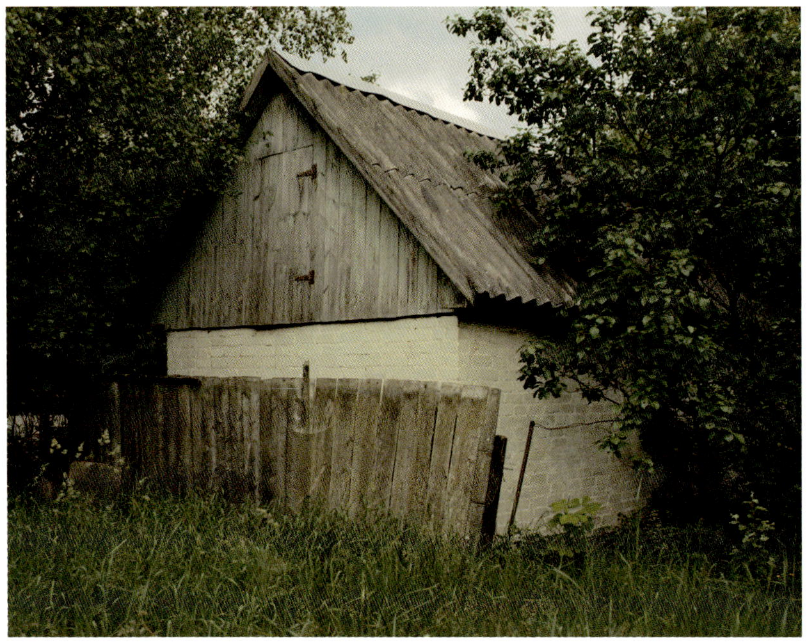
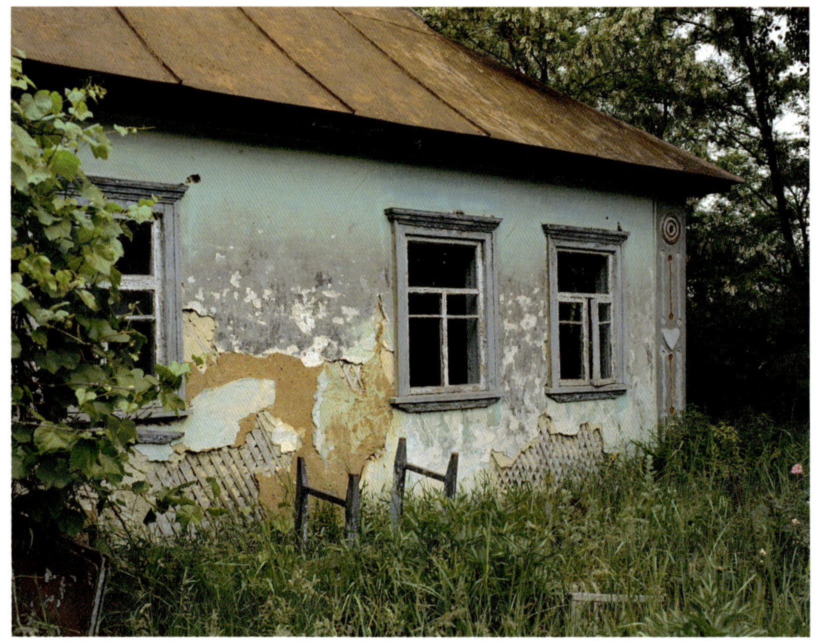

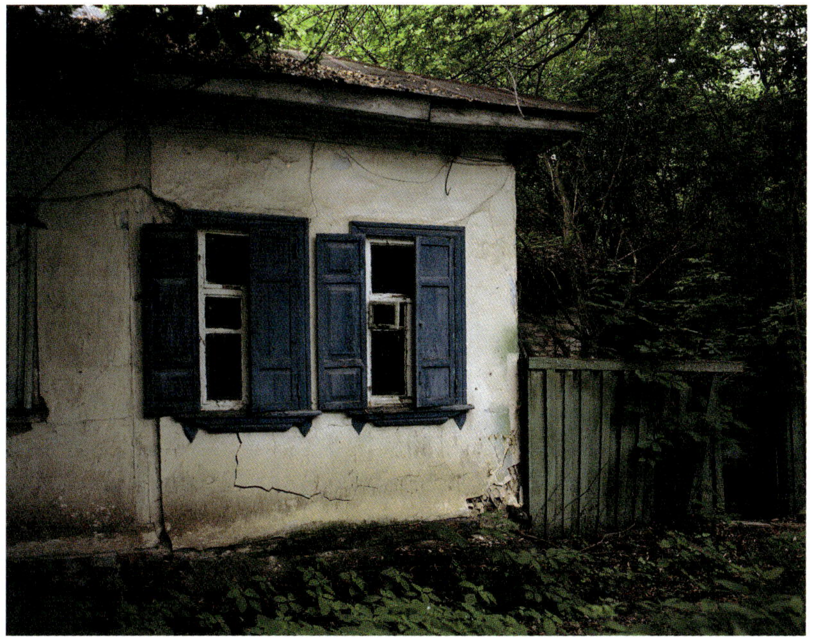

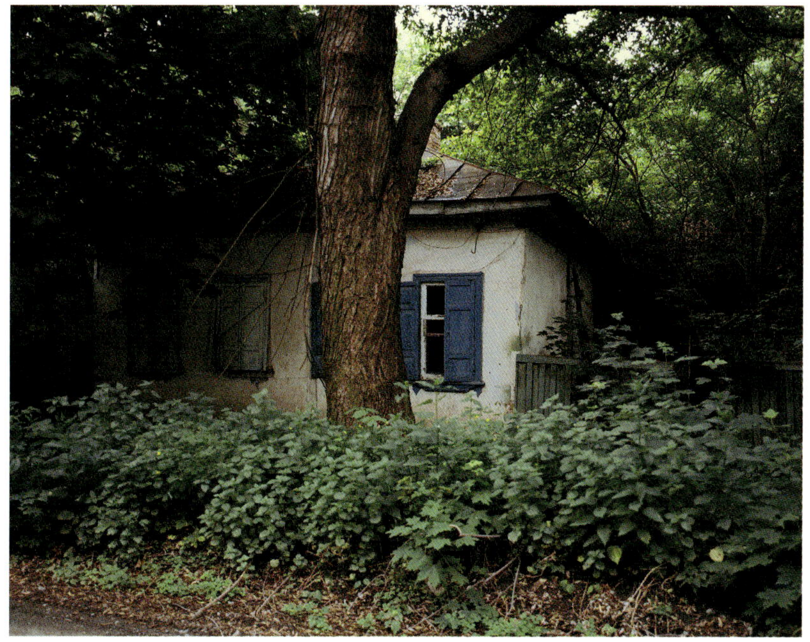

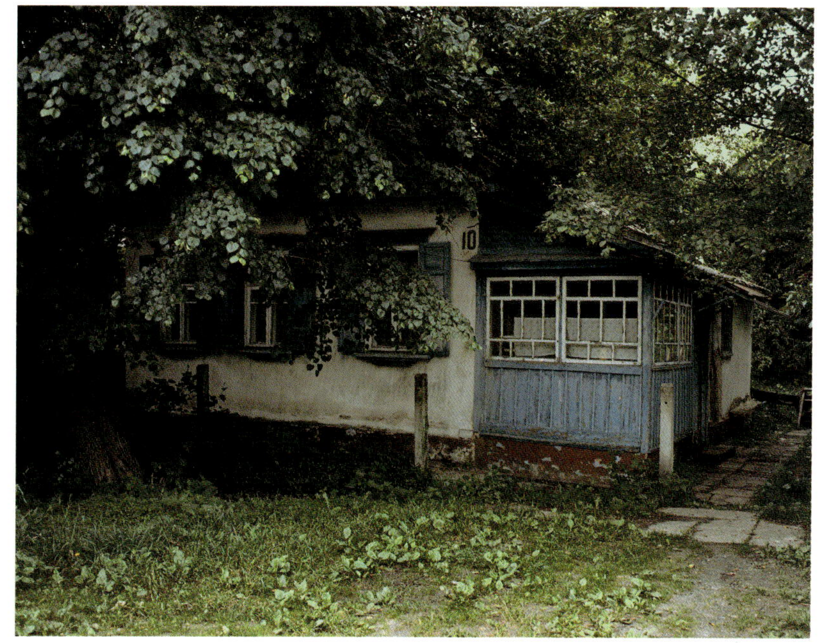
10

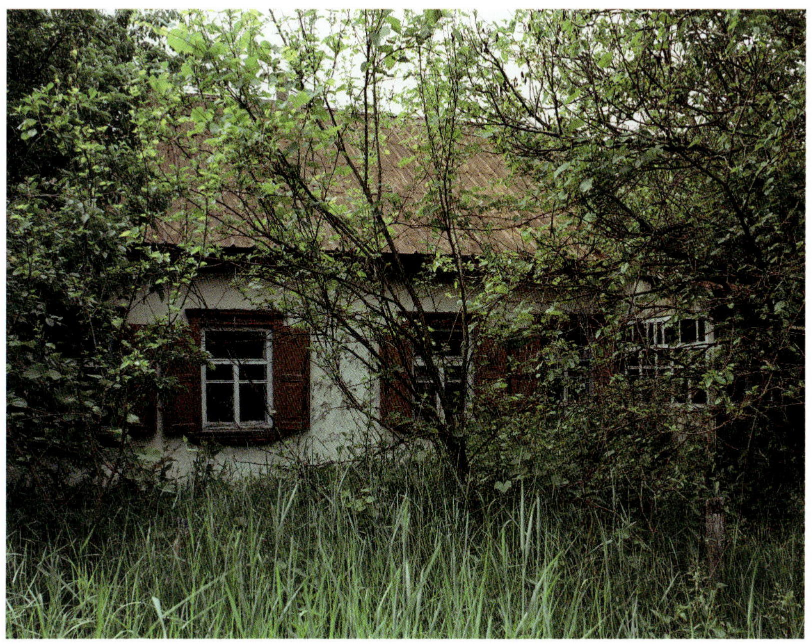

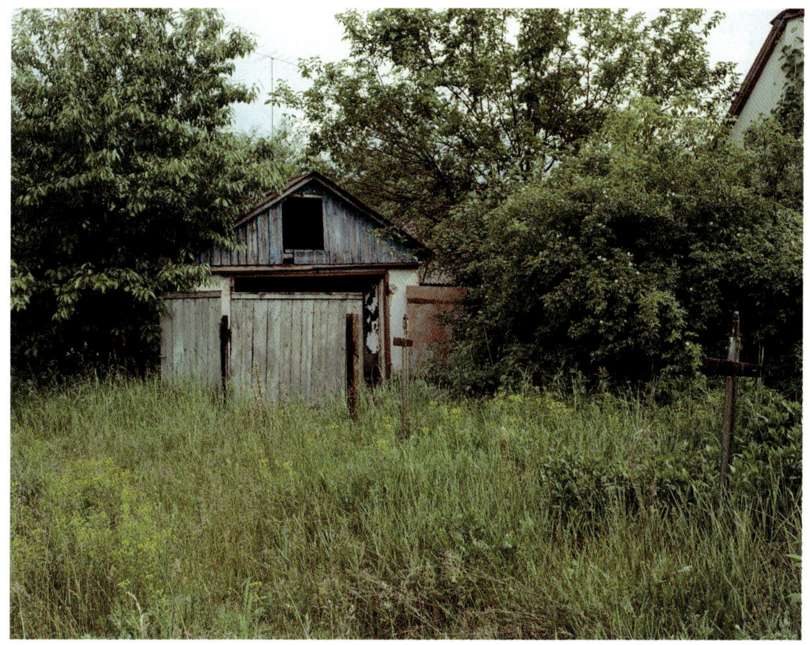

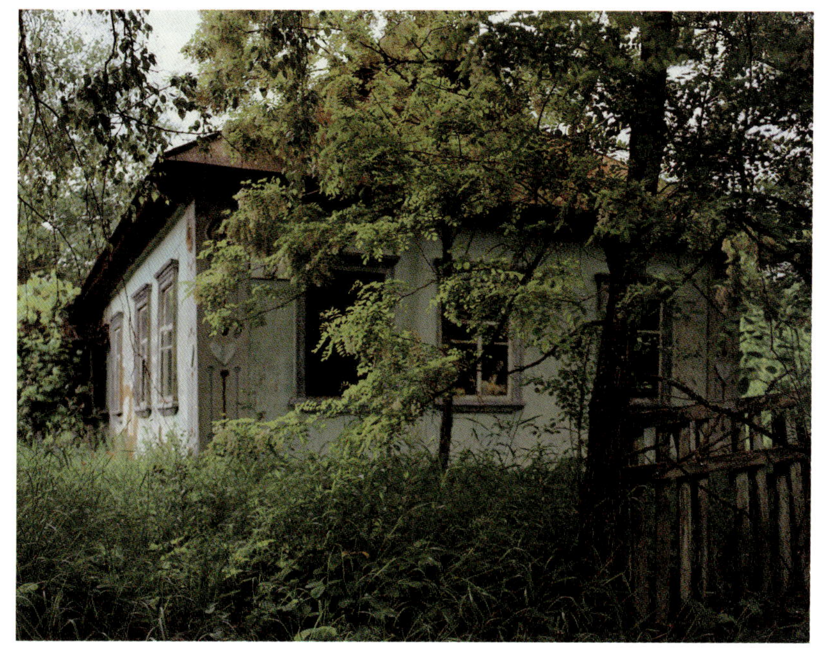

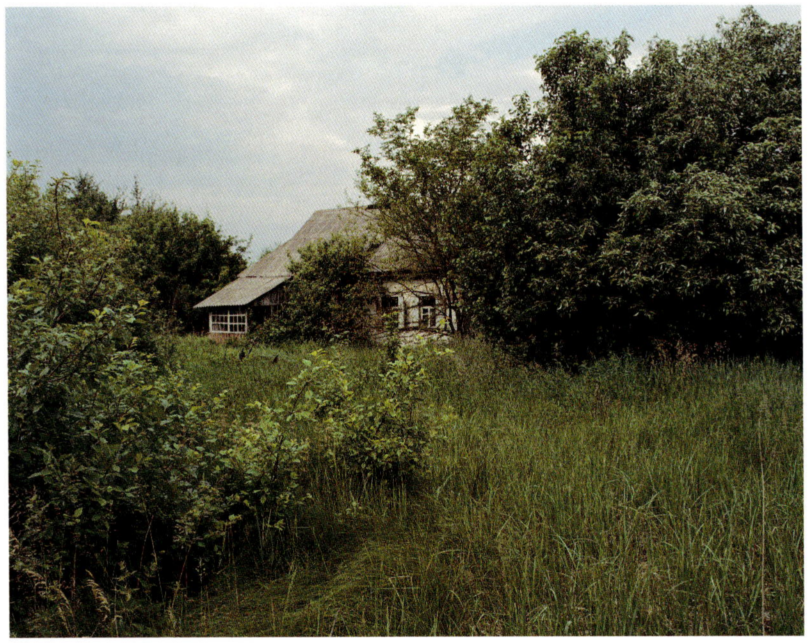
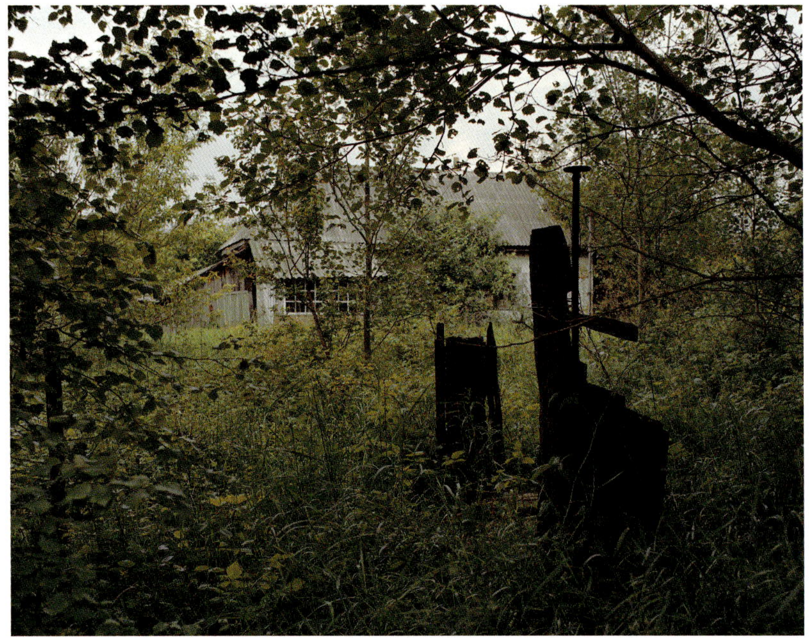
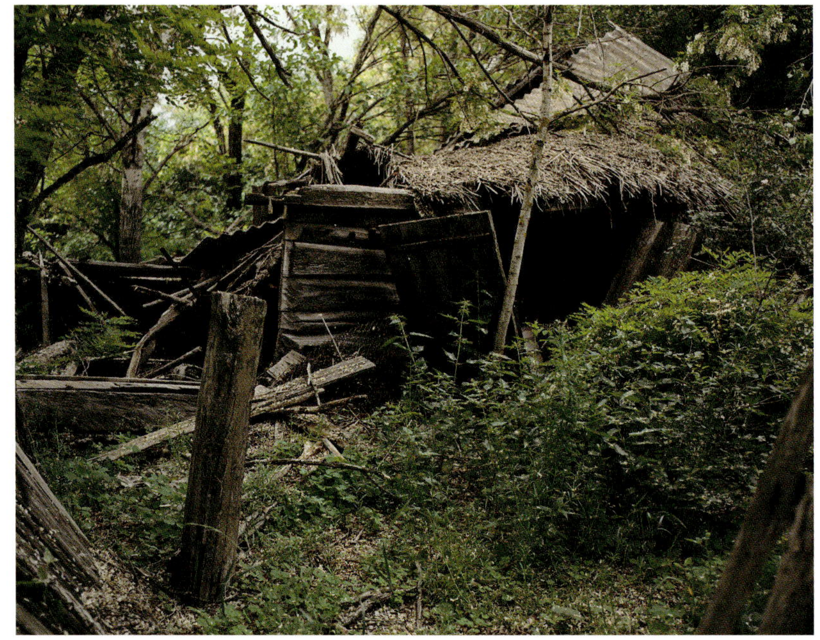
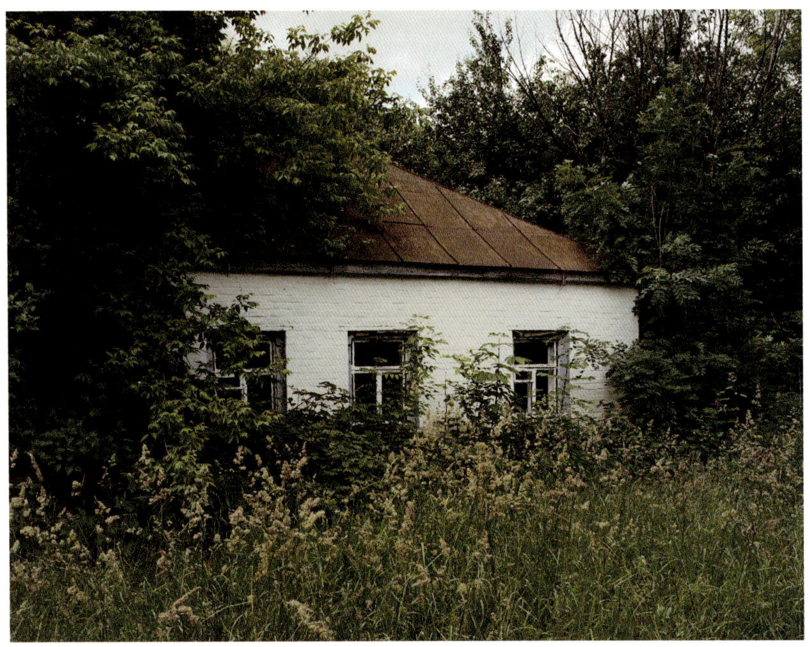
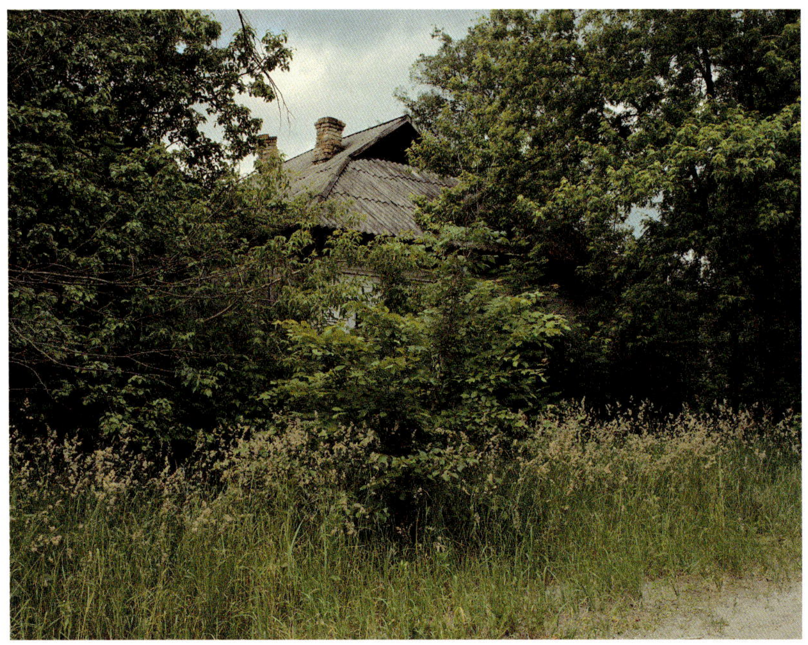
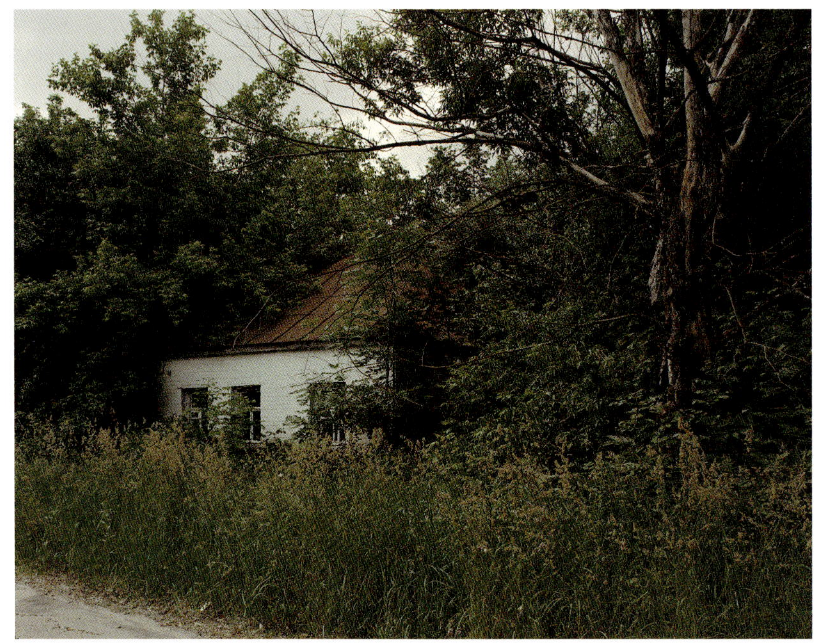

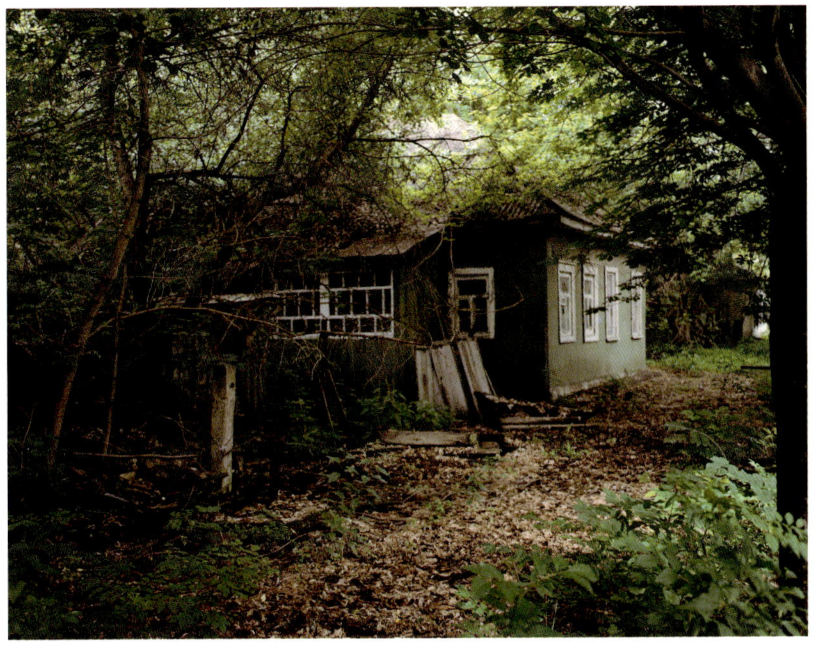
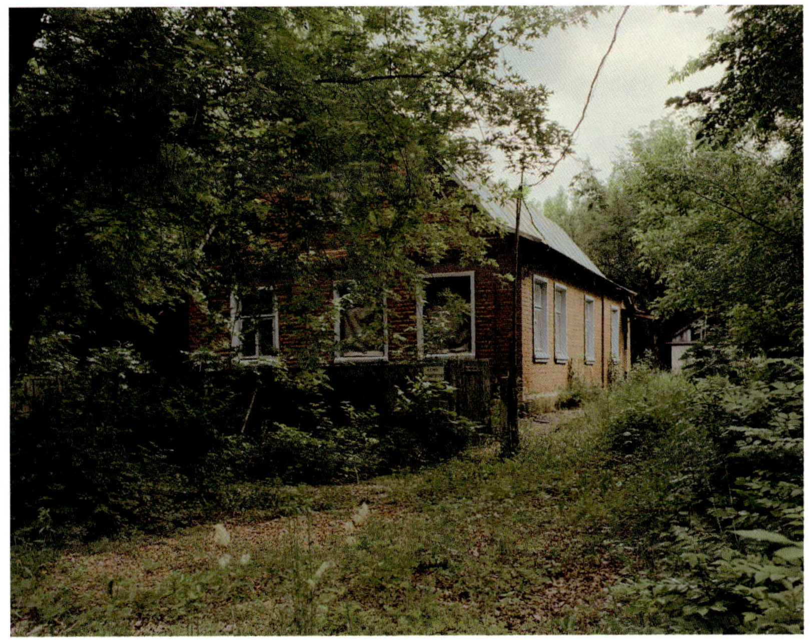
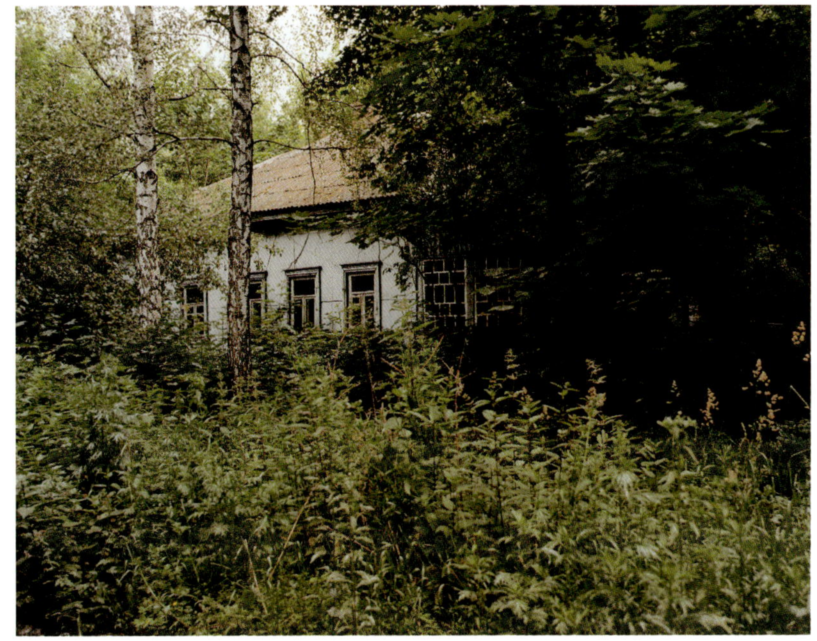
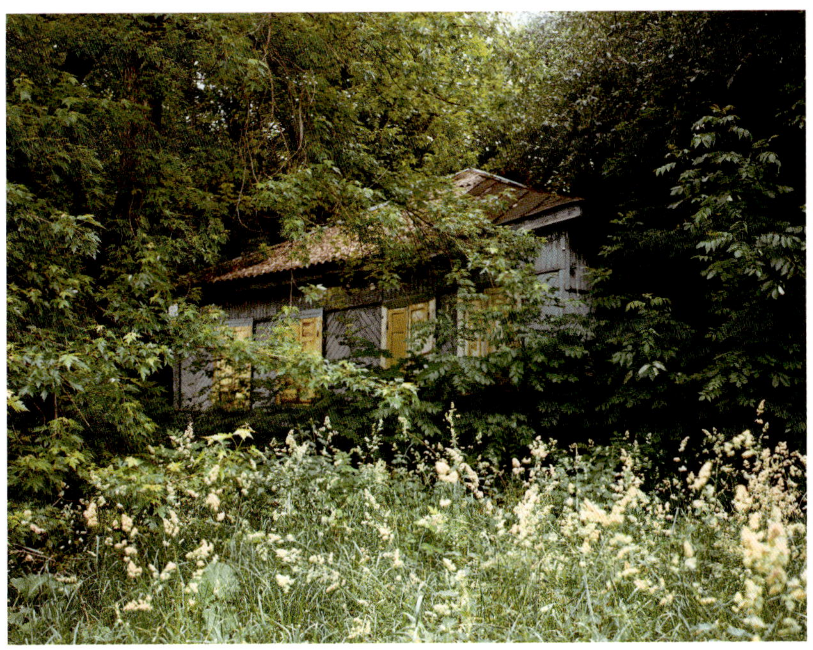
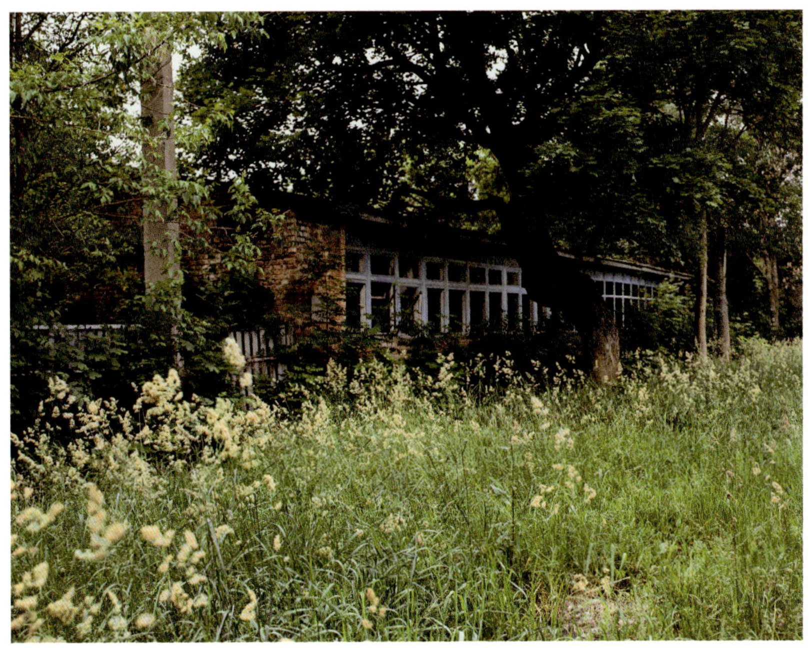
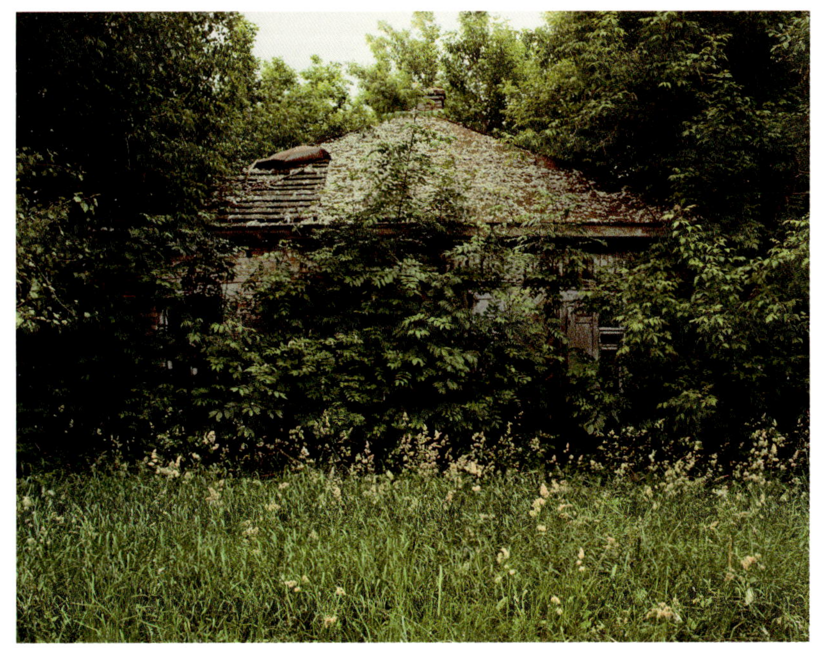

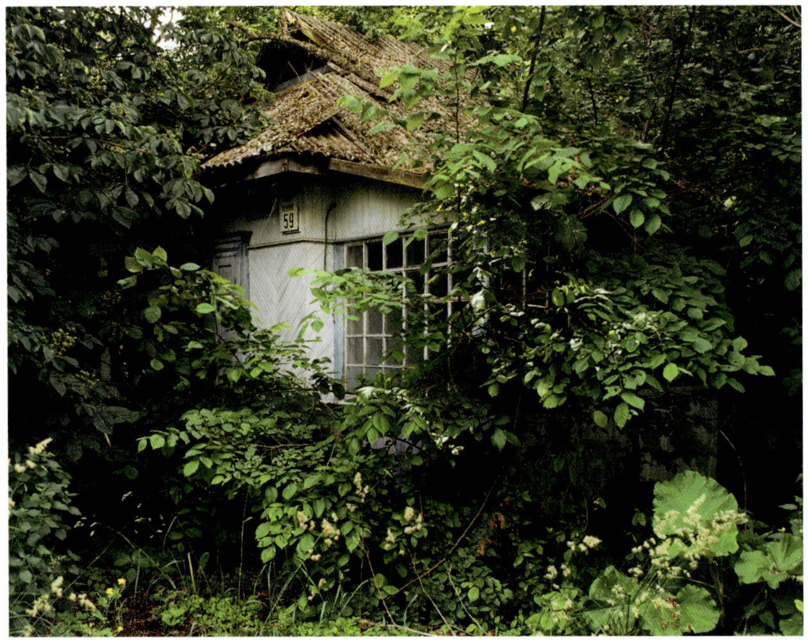

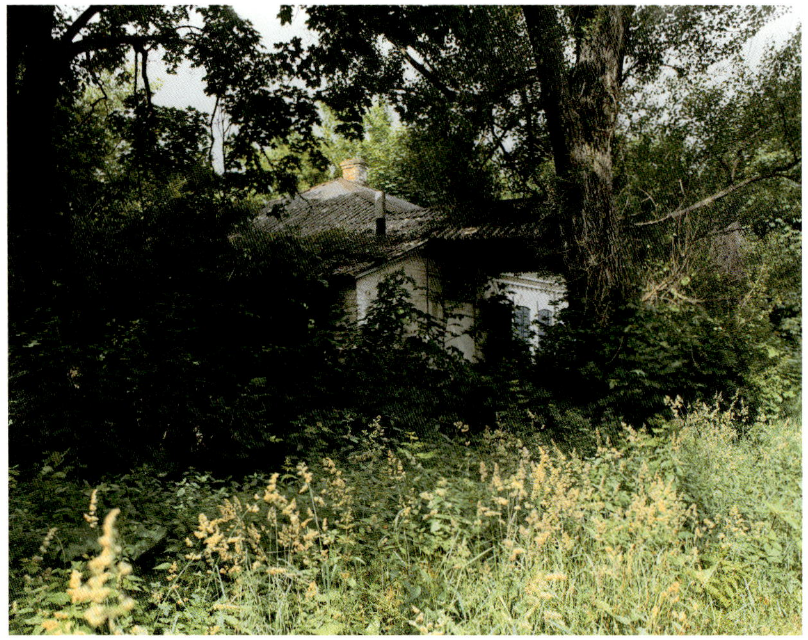

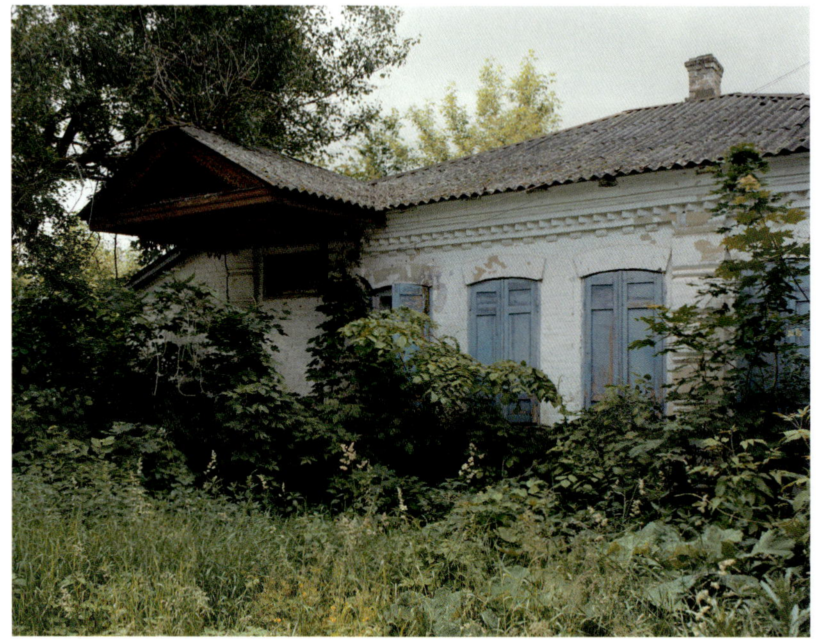

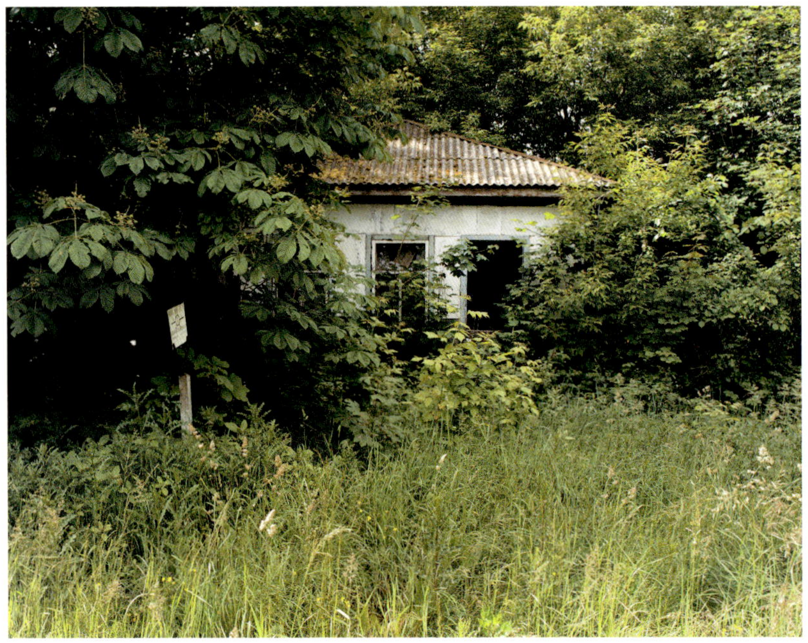

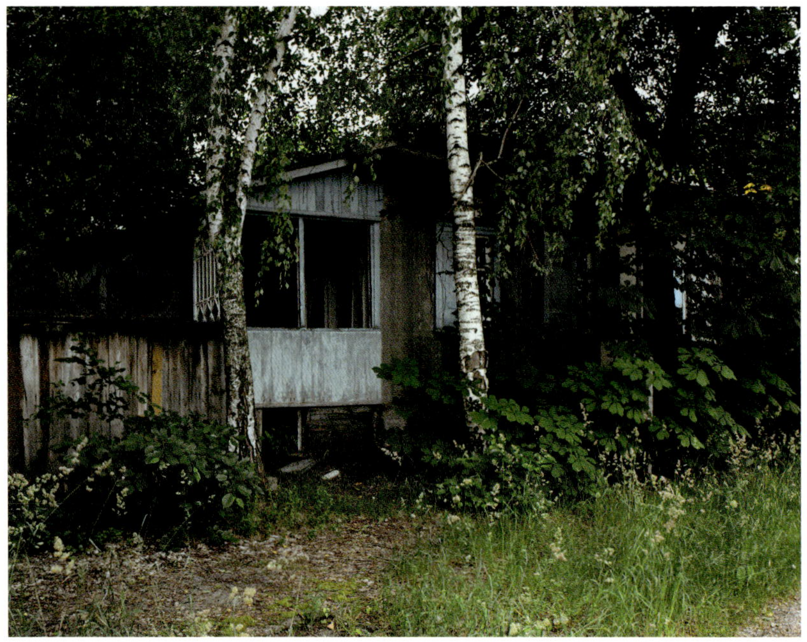

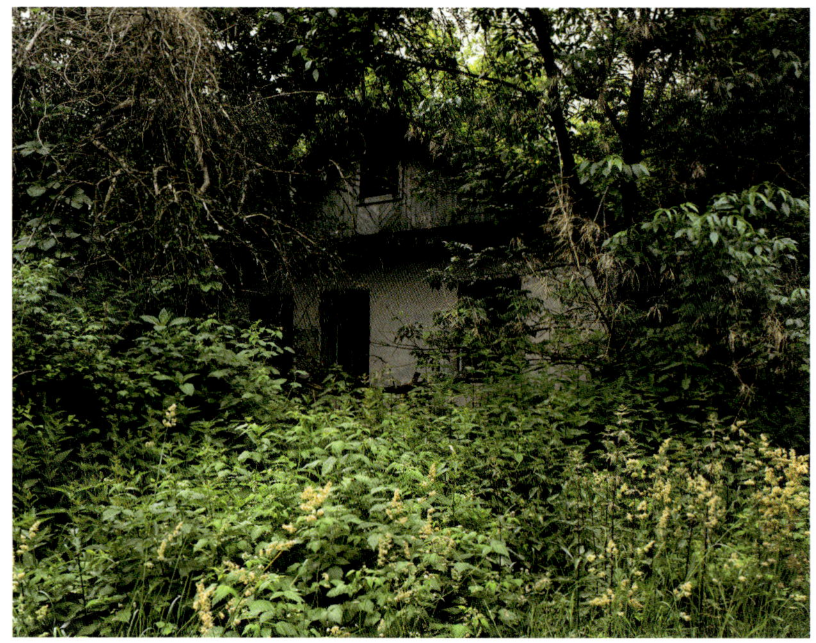

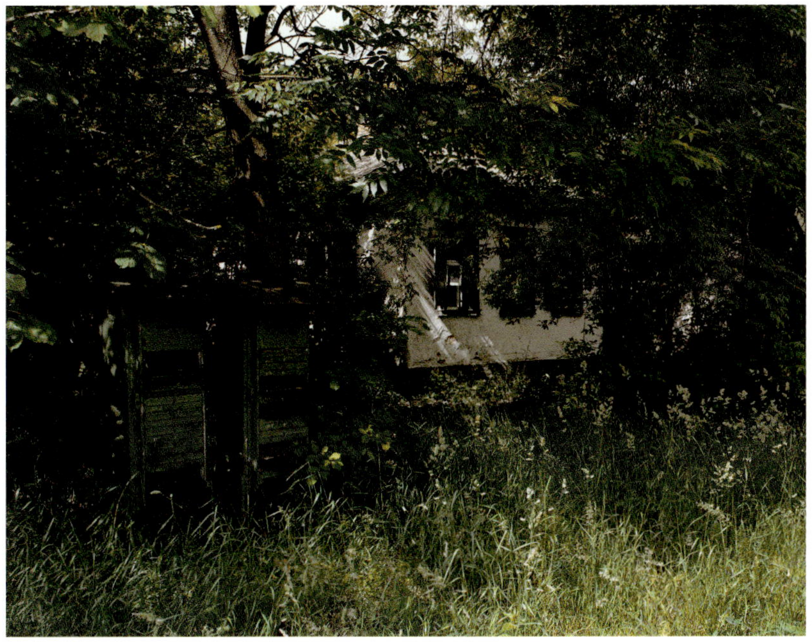
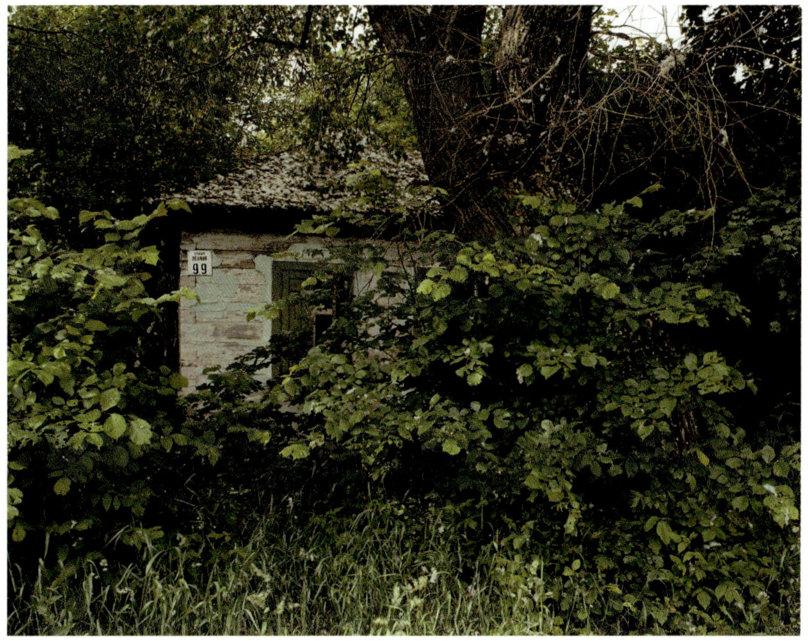
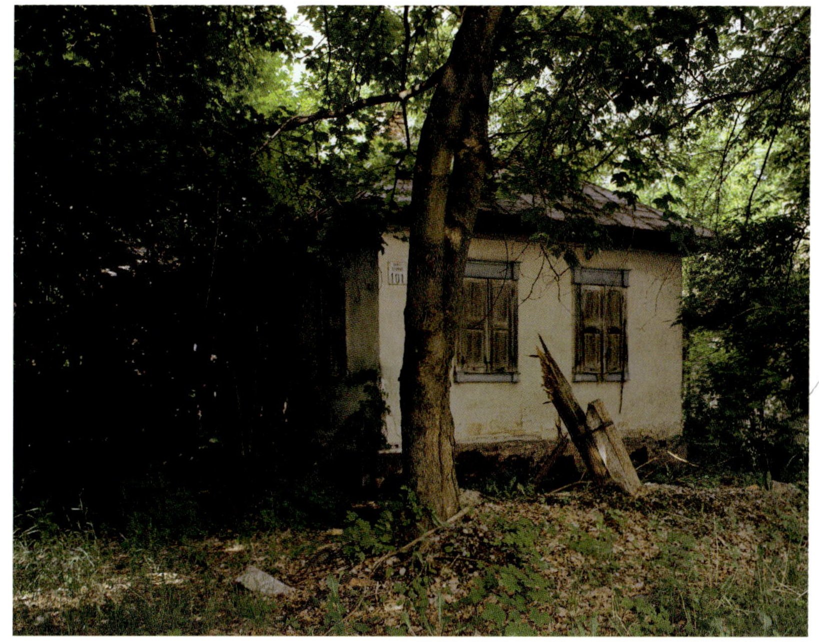
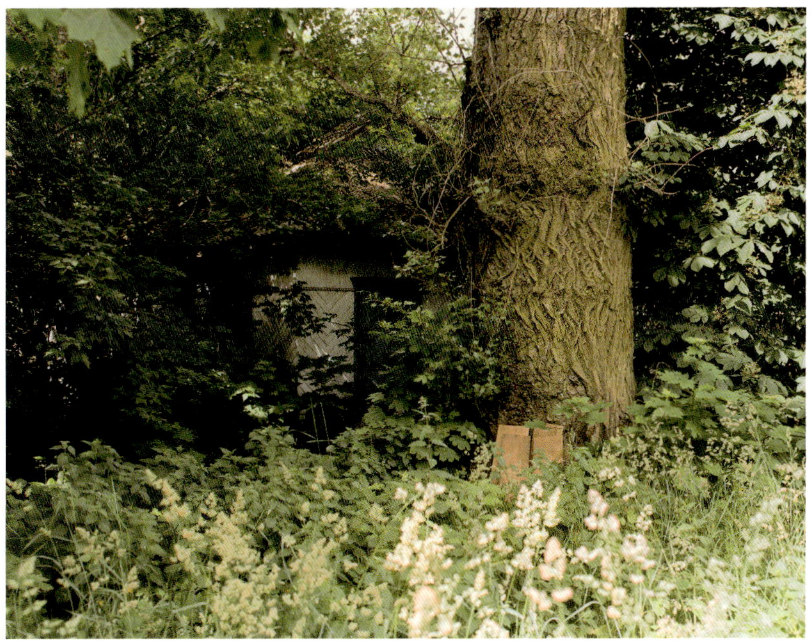
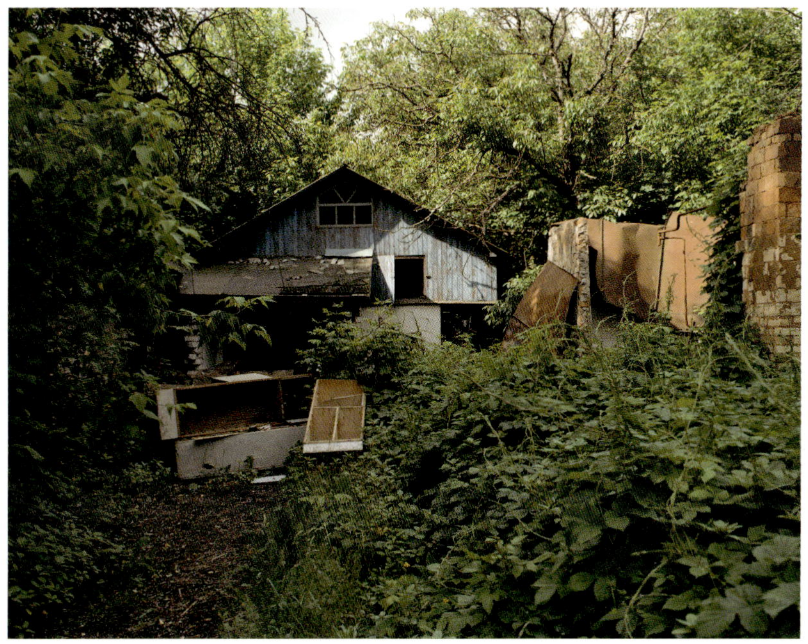
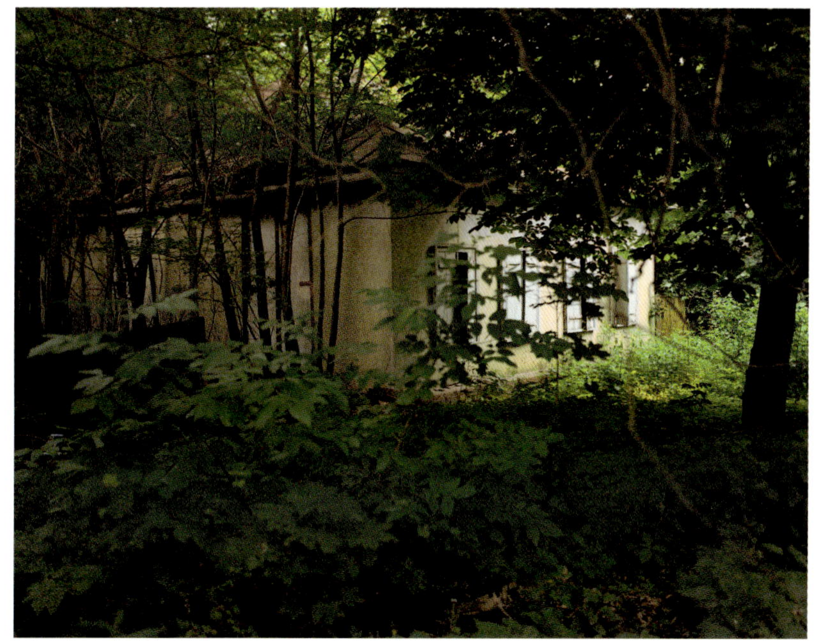

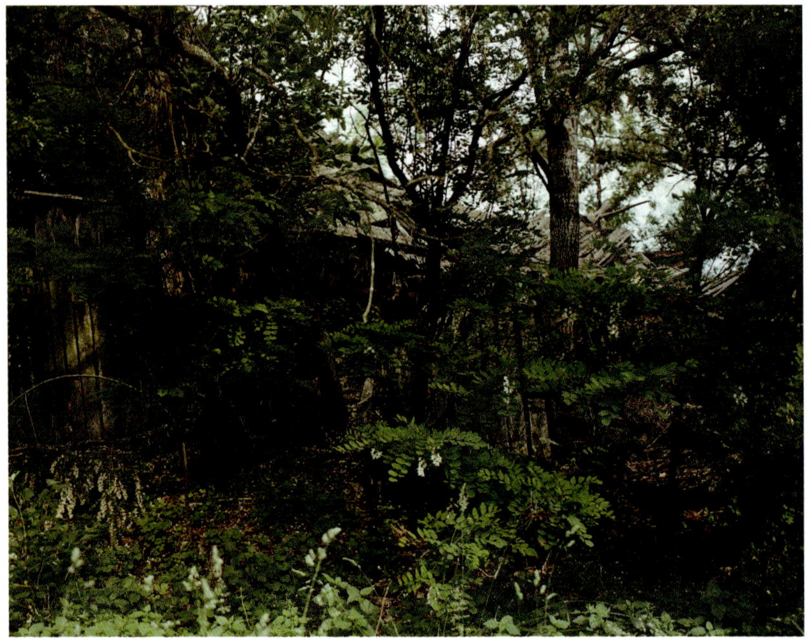
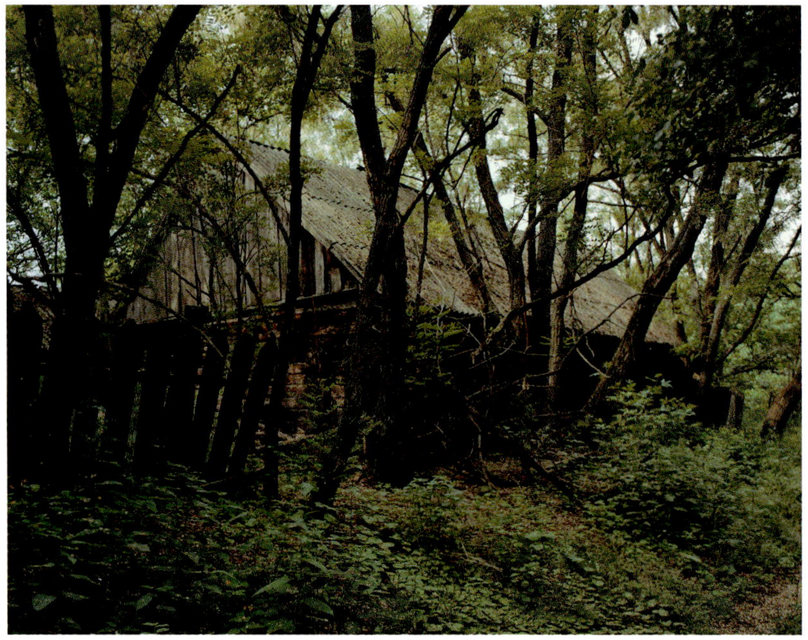
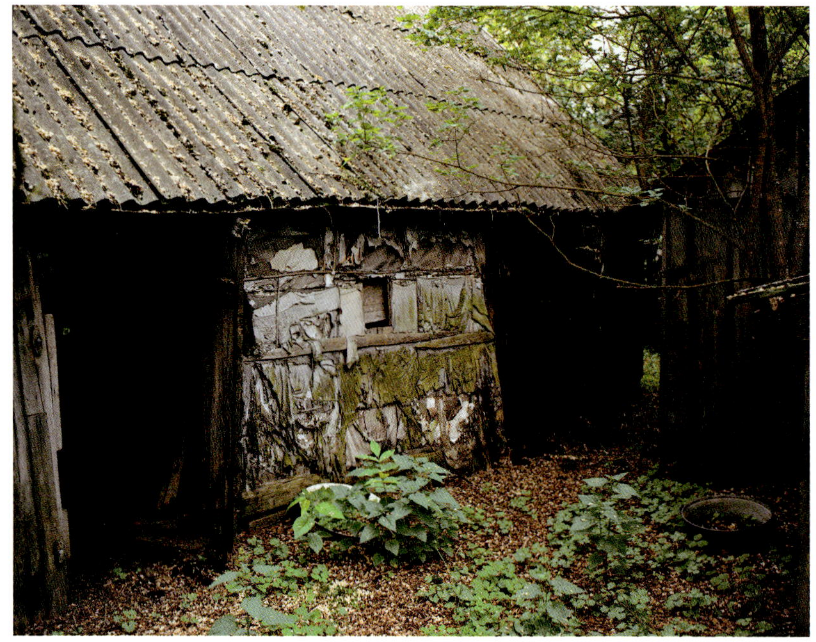
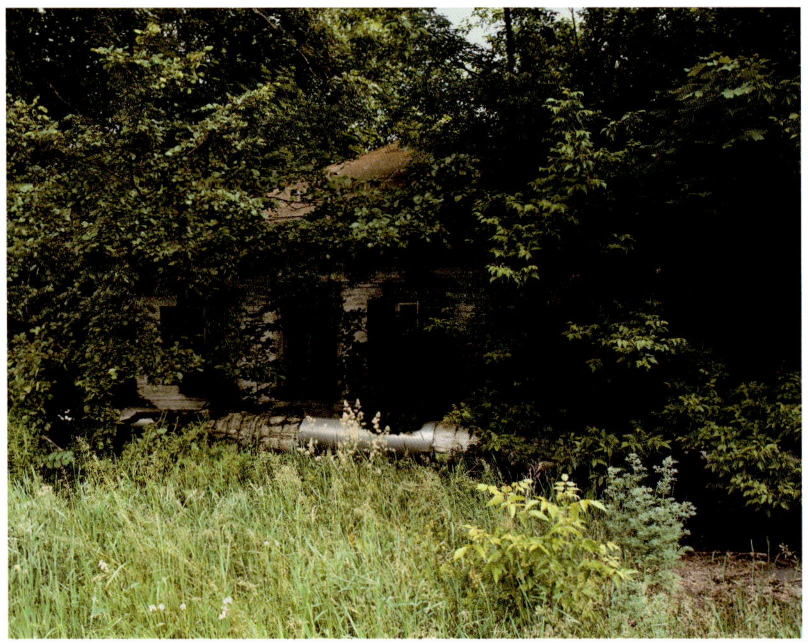
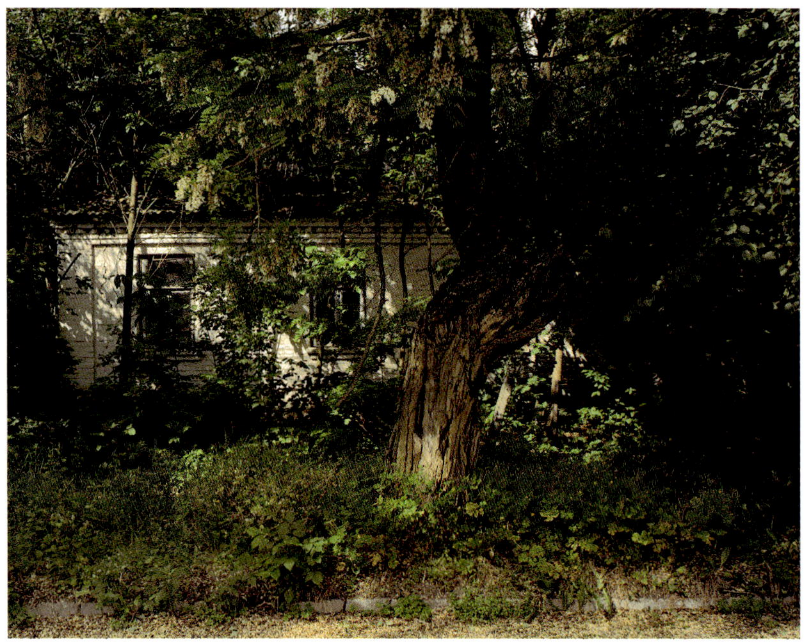
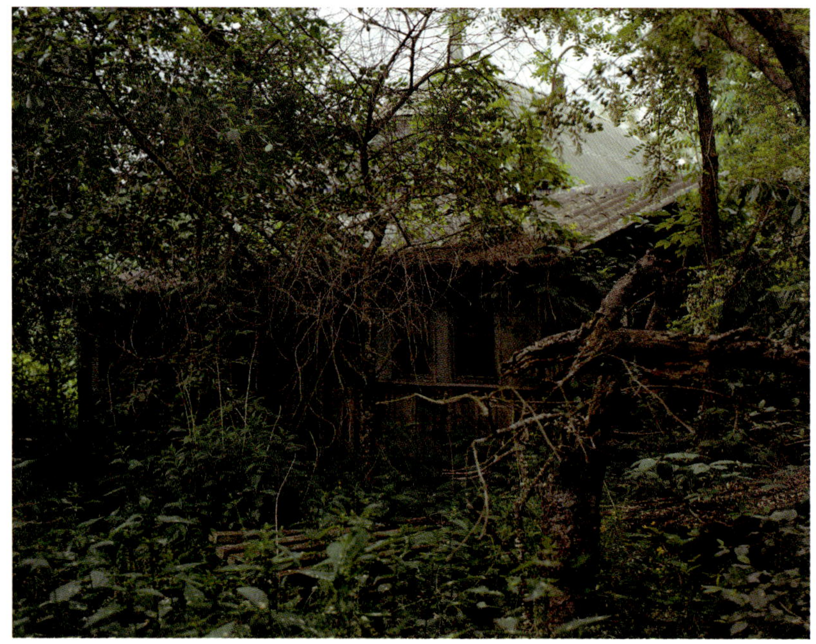

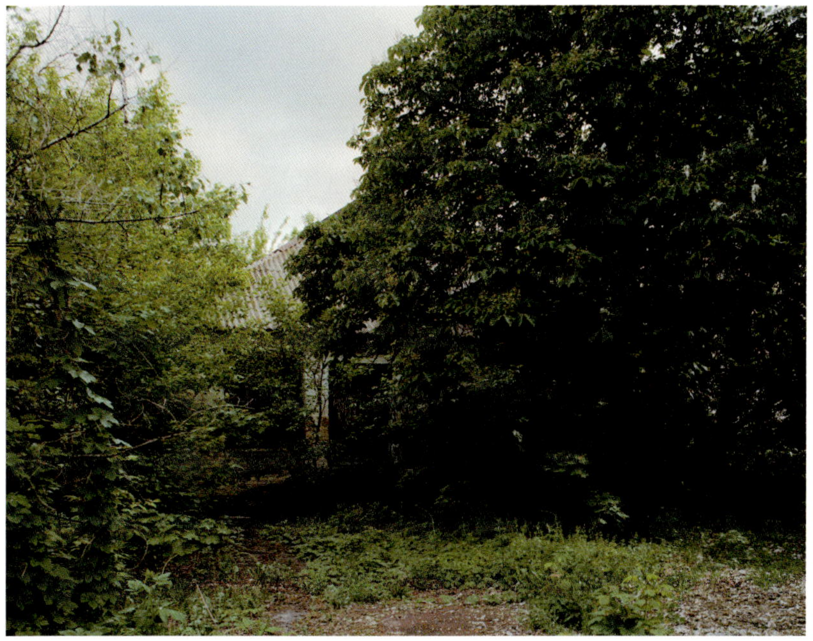
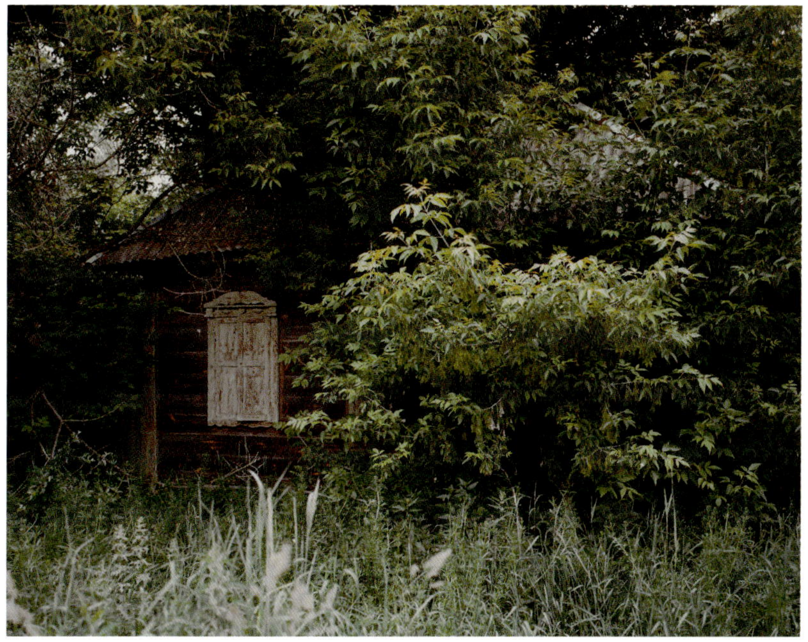
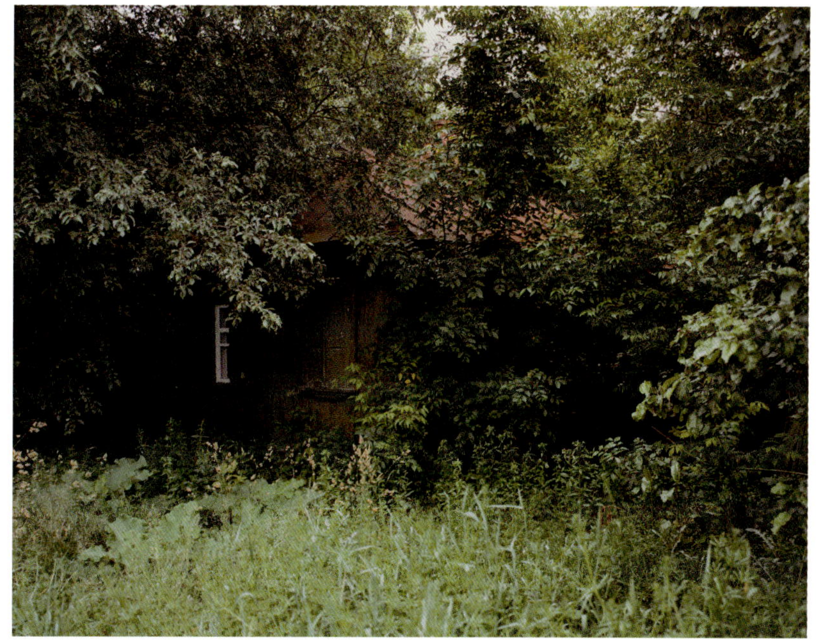
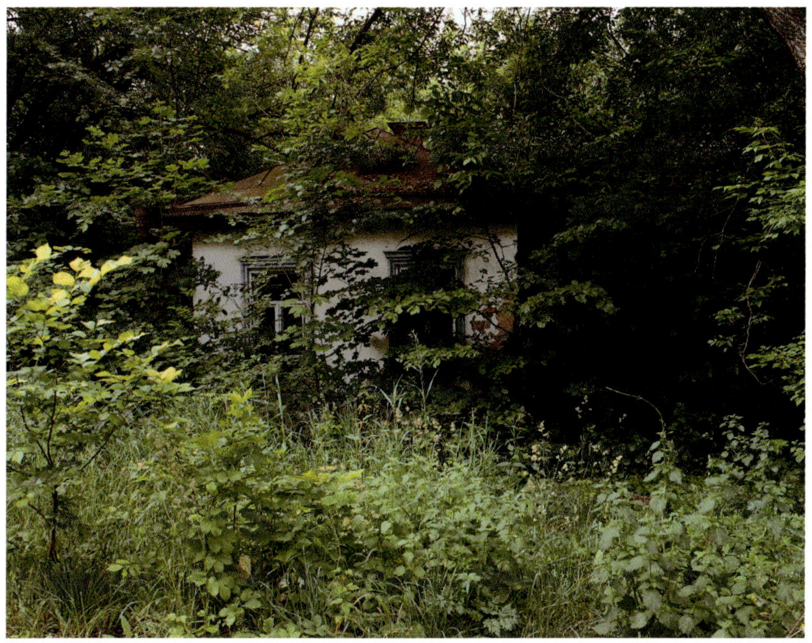
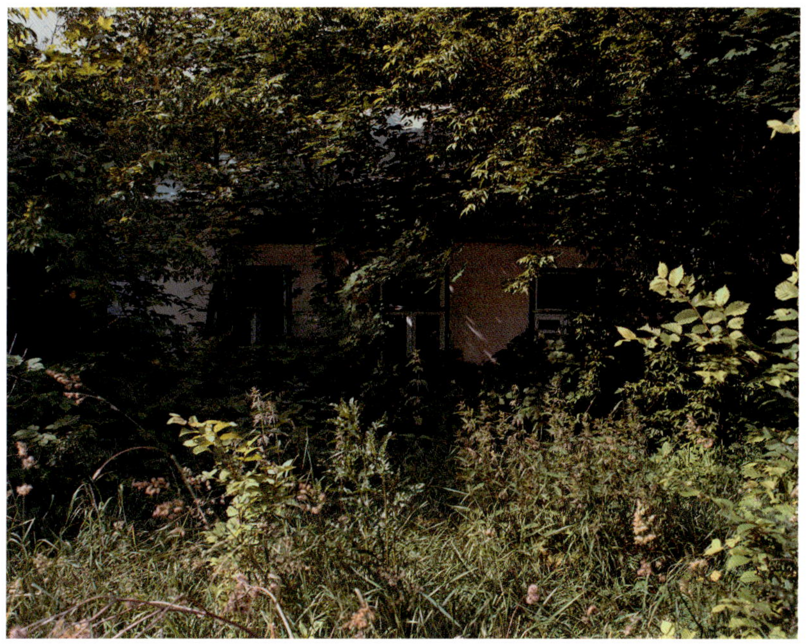
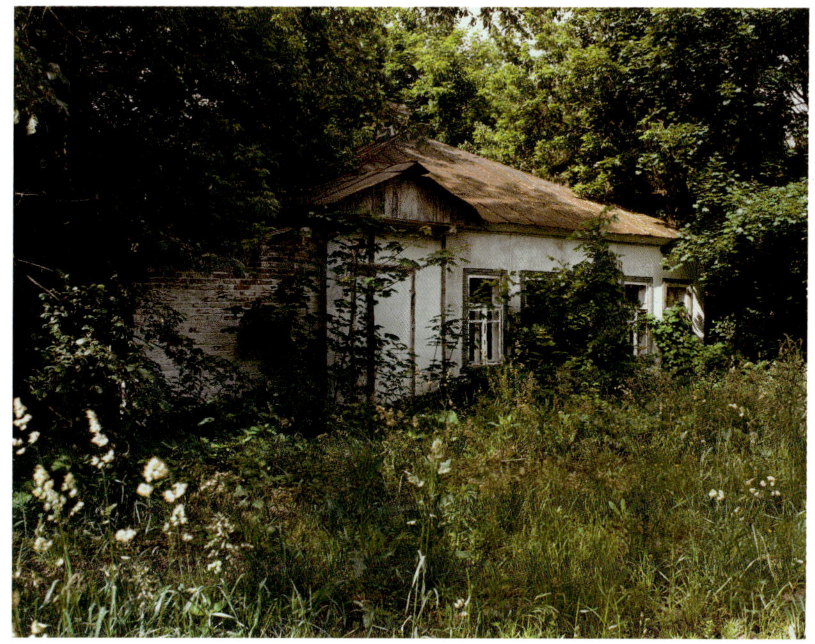

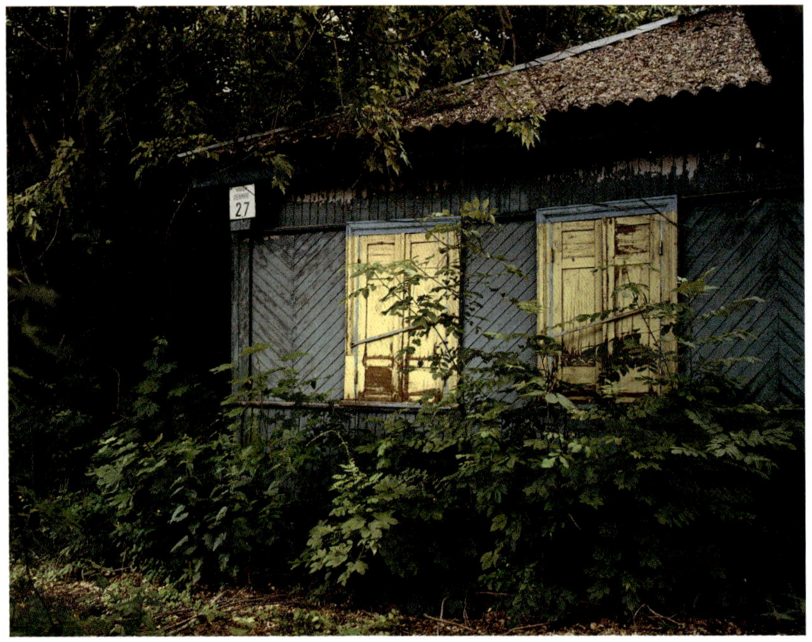
27
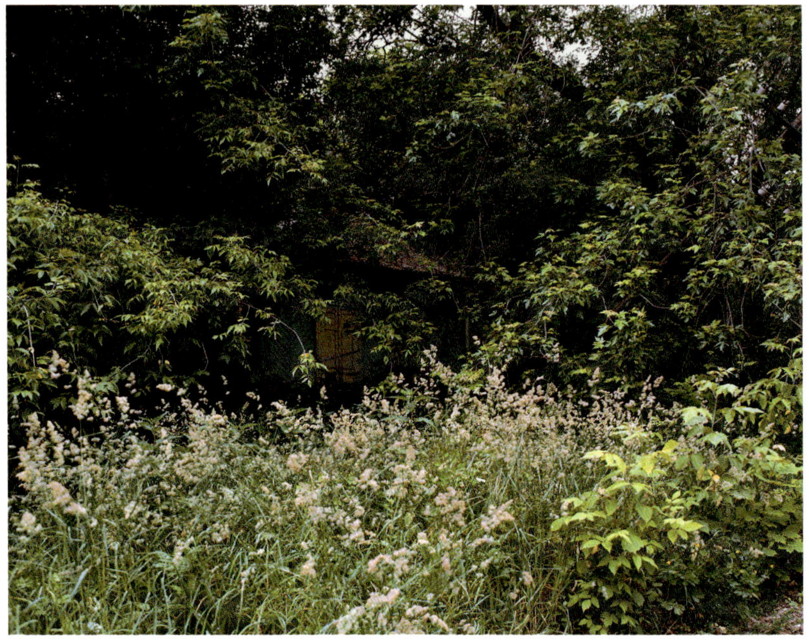
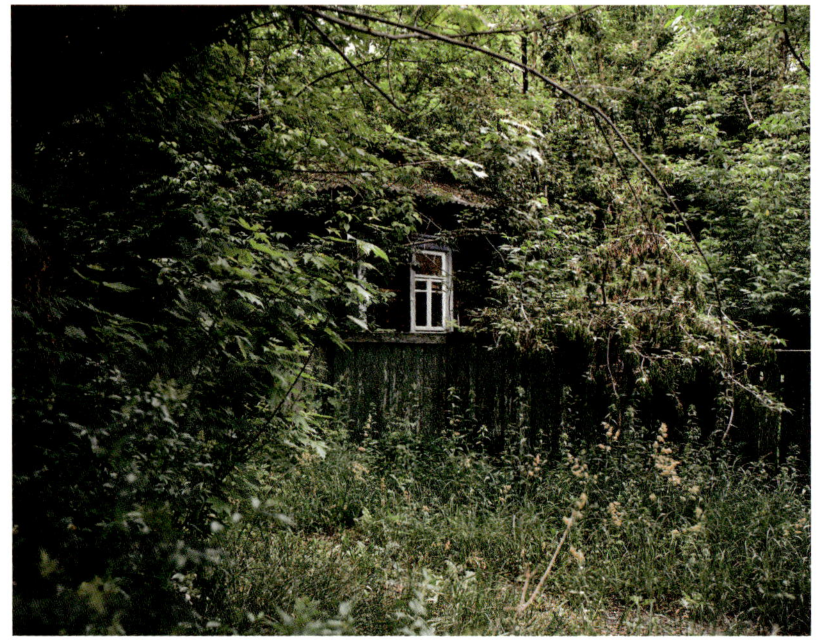
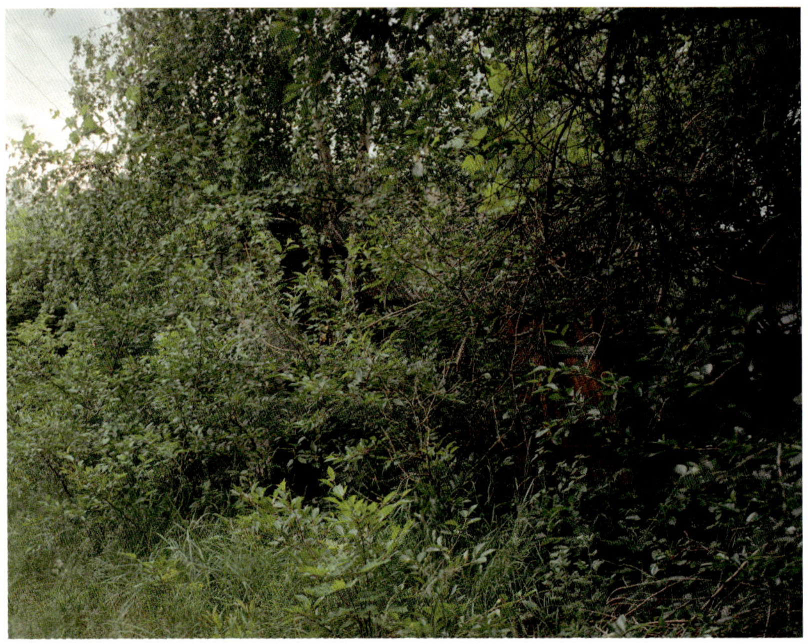
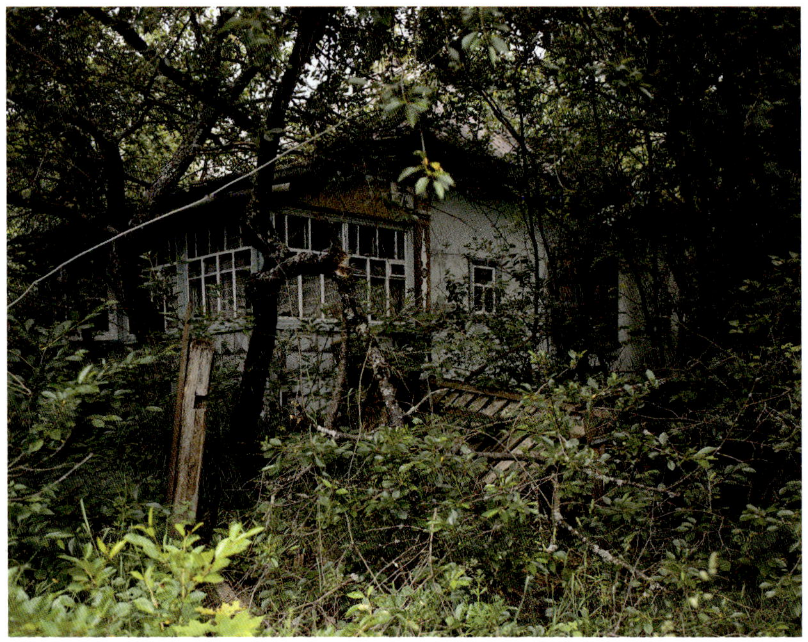
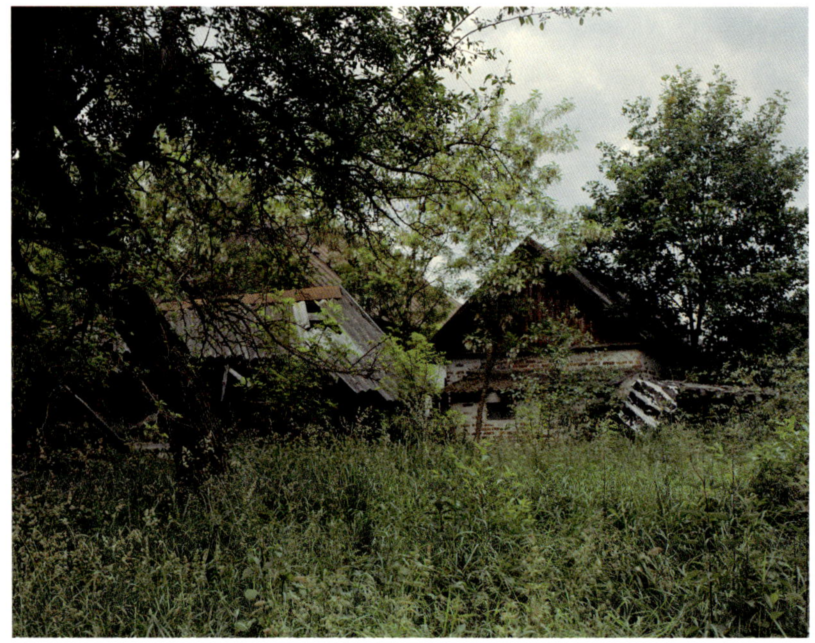

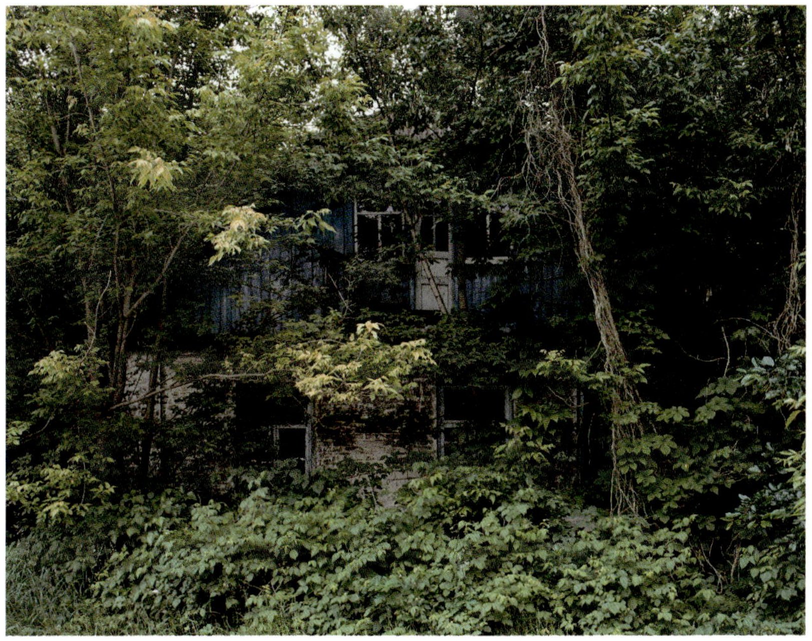
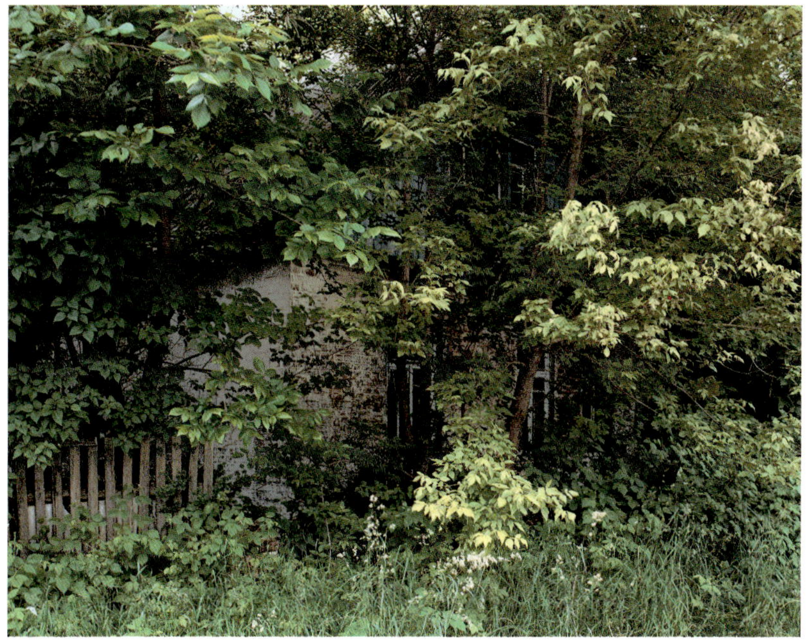
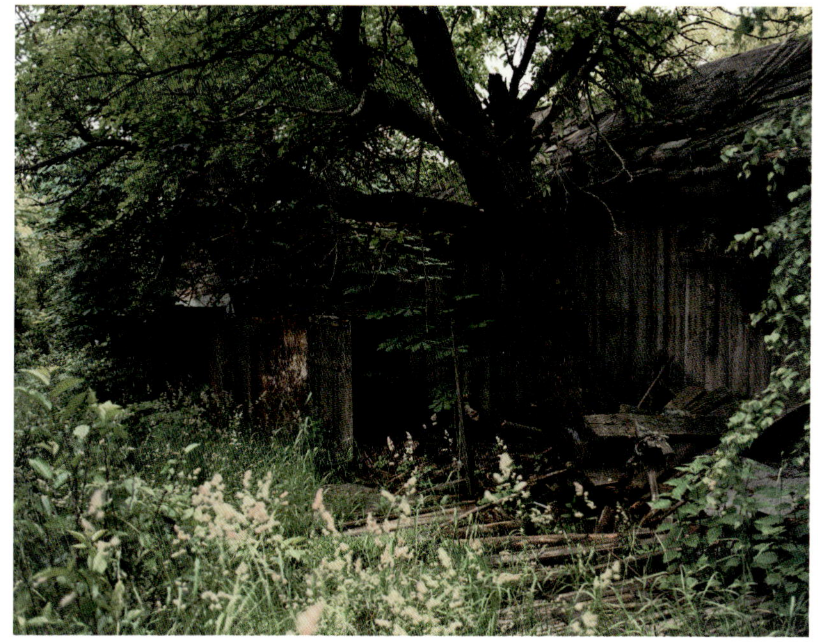
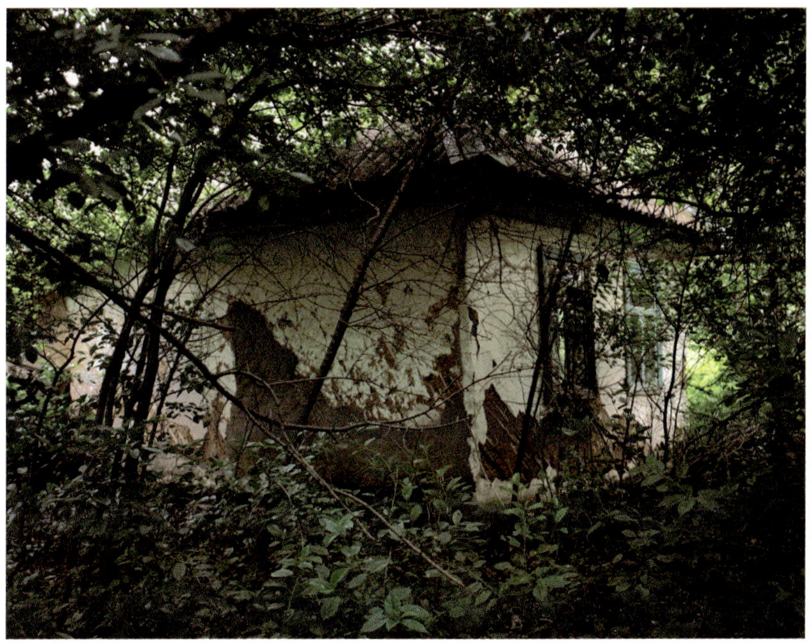
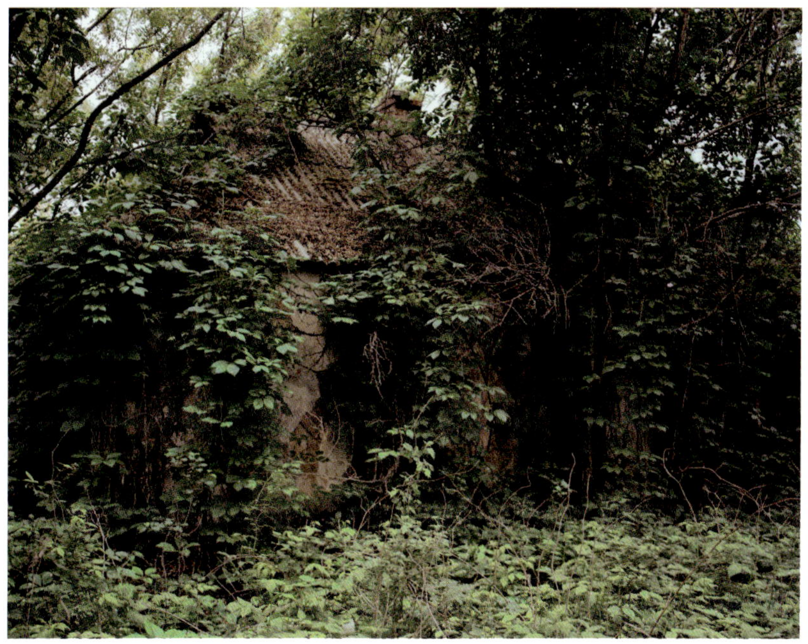
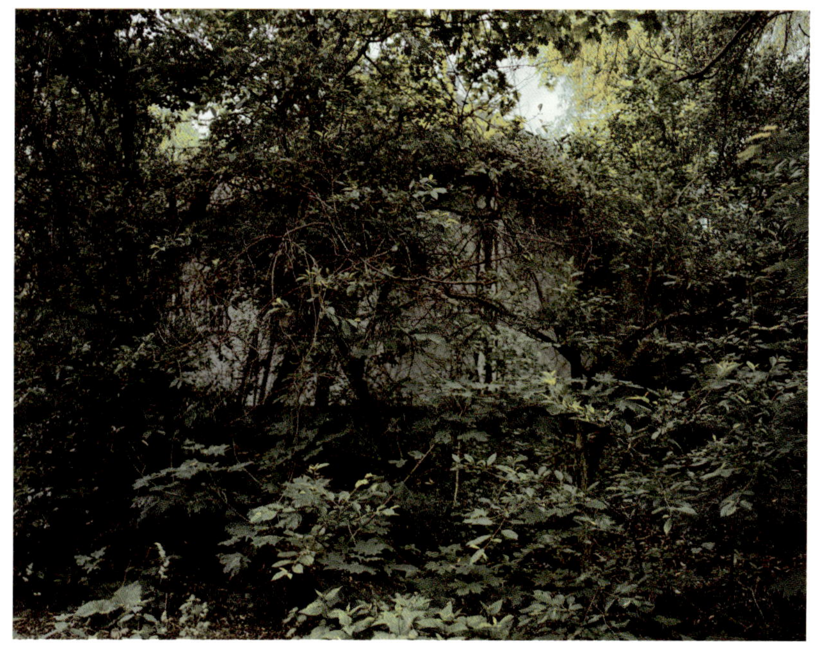

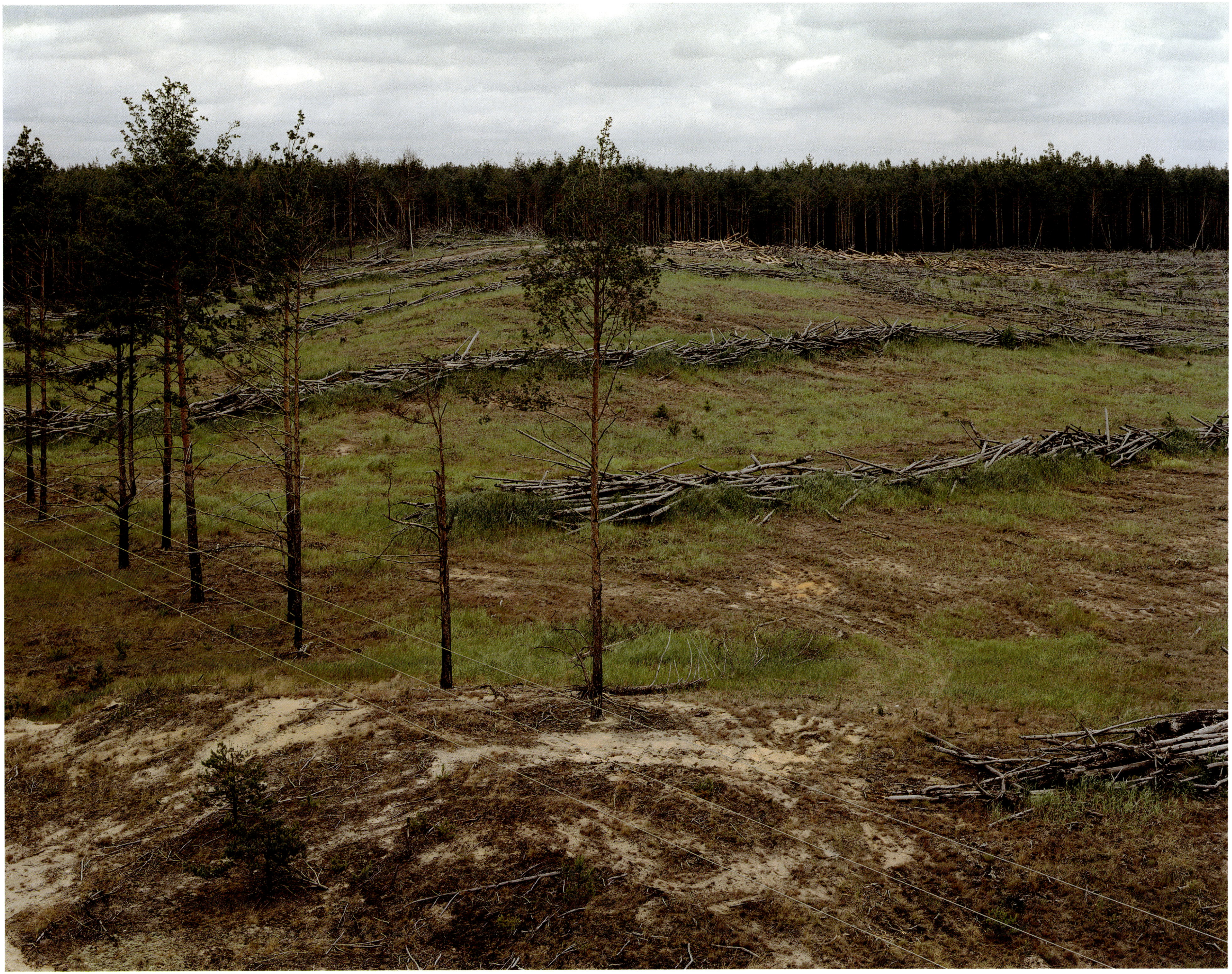

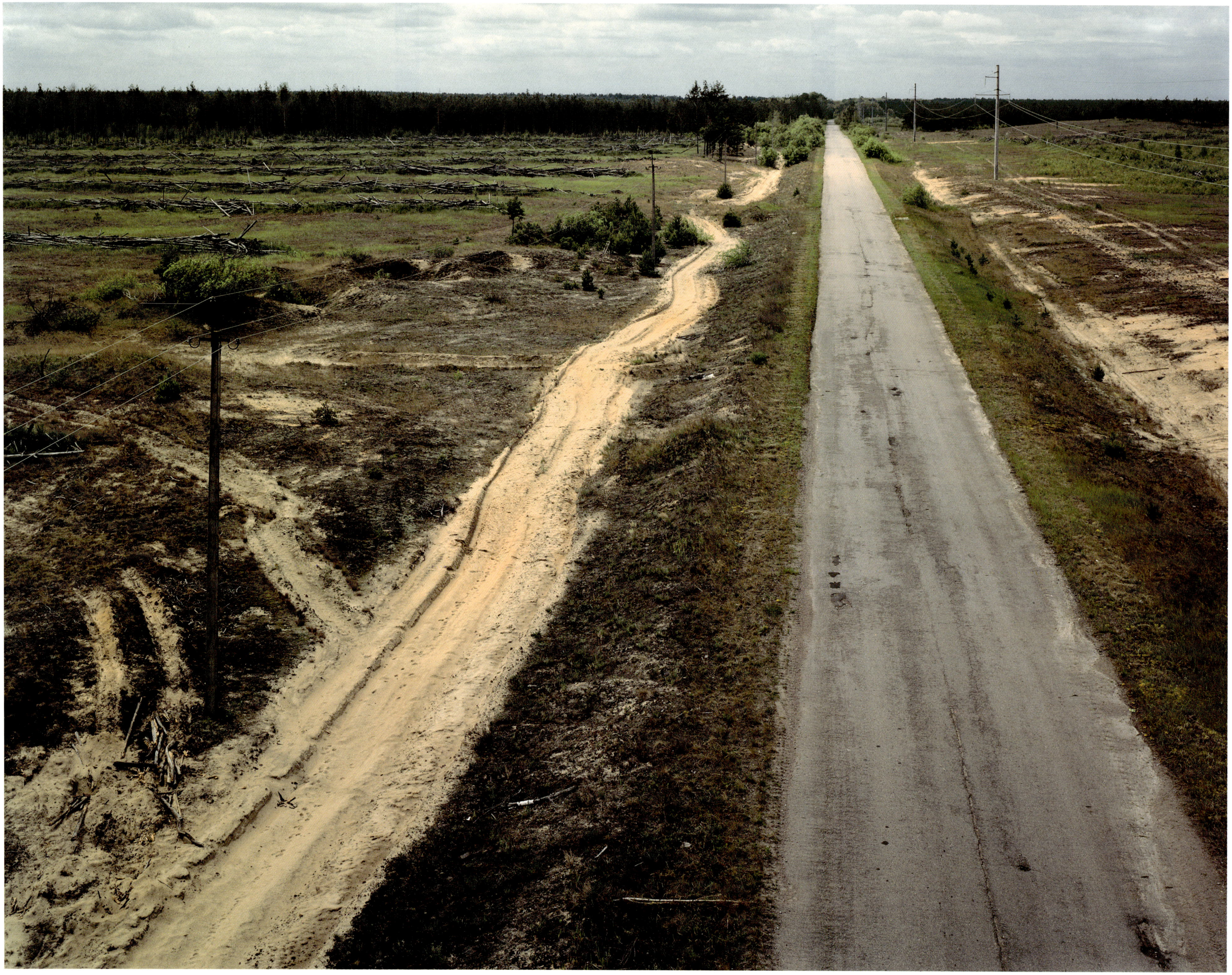

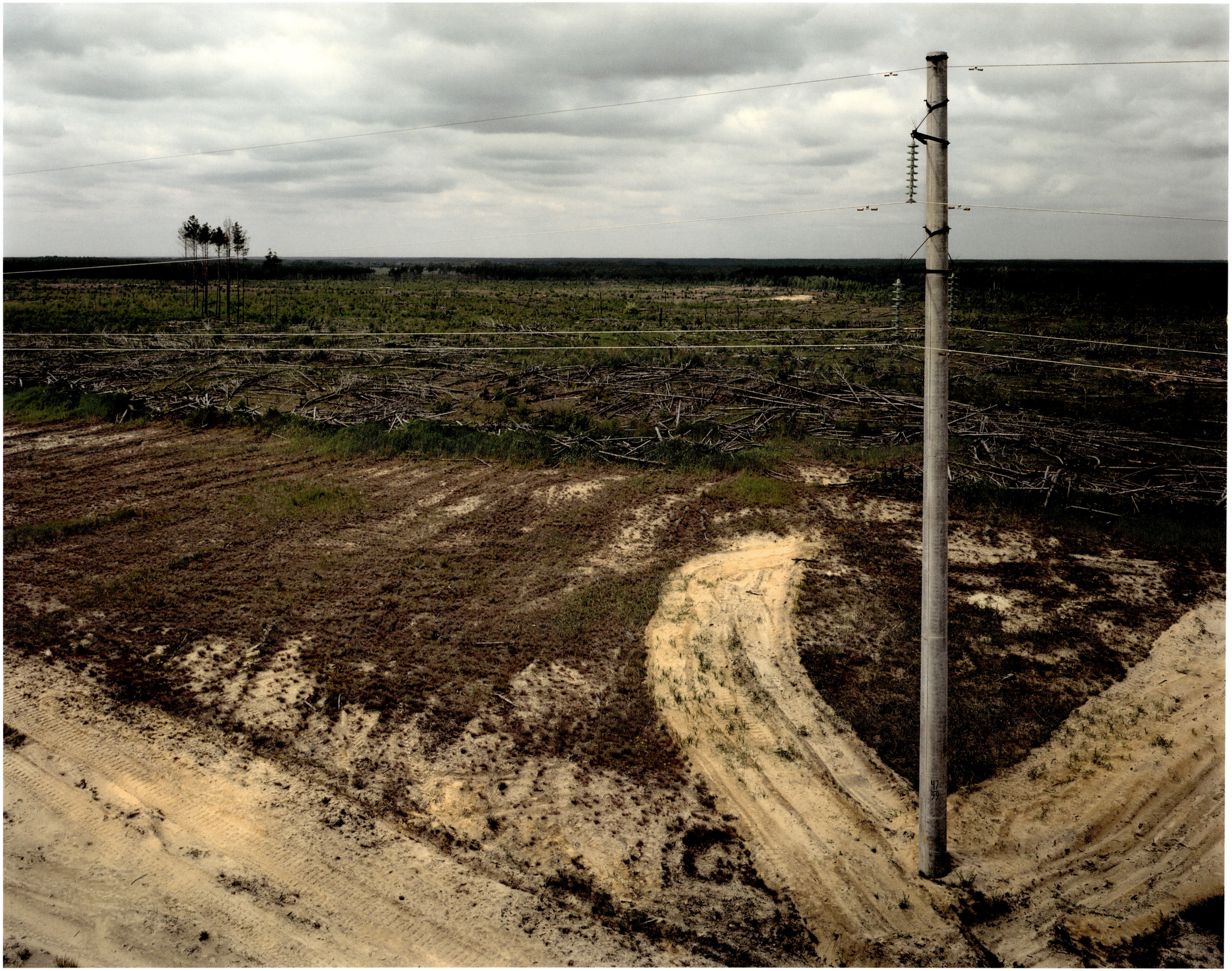

Captions

These photographs were taken June 6-9, 2001.

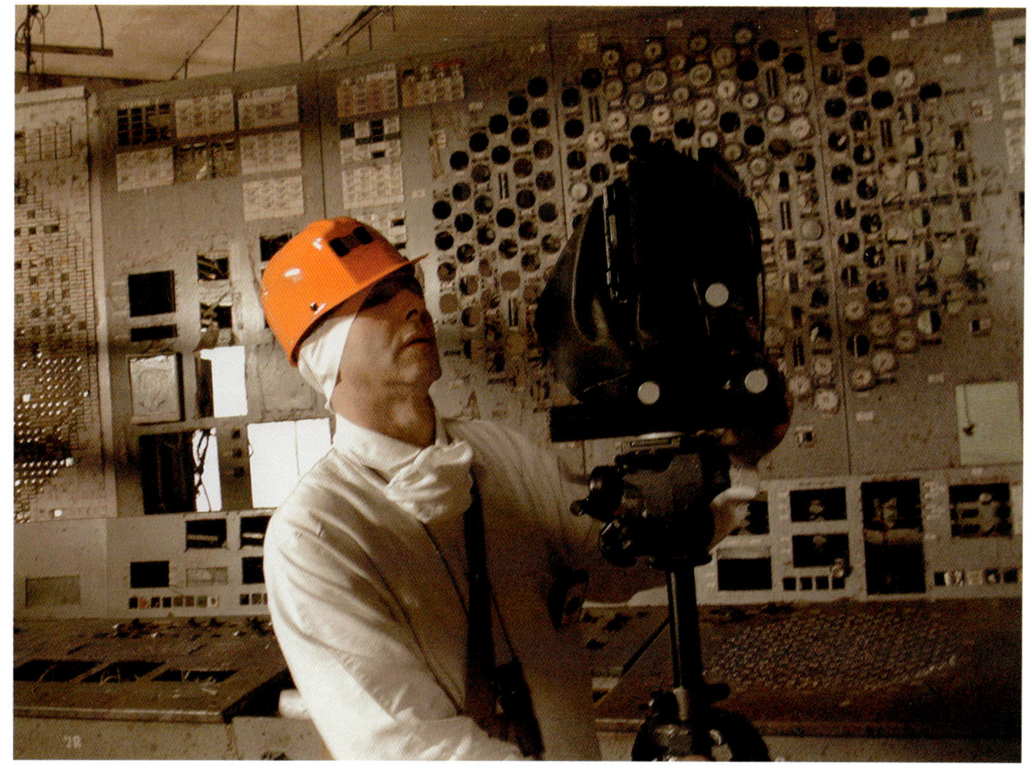

Robert Polidori in the Unit 4 control room. Photograph by Konstantin Leifer, June 6, 2001.

As time passes, our memory of the Chernobyl catastrophe begins to fade. Our initial shock and fear are pushed aside by new world events and private dramas that populate the attention of each of our lives. The radioactive half-lives of elements, however, are not subject to this form of accelerated, subjective amnesia.

Does any generation have the right to risk the safety of so many future generations? It is my belief that this is the fundamental moral question raised by the disaster. I felt personally compelled to confront and witness this ongoing tragedy that no ritual can heal.

I thank my friends and family for their support of this project. In particular, I thank Konstantin Leifer for not only securing necessary visas and authorizations, but for acting as a guide and companion as I investigated what remains in the Zone of Absolute Resettlement, situated inside the Exclusion Zone around Chernobyl in Ukraine.

Robert Polidori
Göttingen, April 2003

First edition 2003
Second edition 2004

Photography: Robert Polidori

Book design: Robert Polidori, Elizabeth Culbert and Gerhard Steidl
Production: Bernard Fischer
Scan operator: Stefan Hartmann
Color corrections: Katja Töpfer, Judith Lange and Inès Schumann
Typography: Bernard Fischer / Claas Möller
Maquette fabrication: Jonas Wettre
Scans done at Steidl's digital darkroom
Printing by Steidl, Göttingen

Distribution: Steidl, Düstere Str. 4, D-37073 Göttingen
phone: +49-551-496060, fax: +49-551-4960649
www.steidl.de E-mail: mail@steidl.de
Orders can be placed directly at our publishing house

ISBN 3-88243-921-1
Printed in Germany